GROUND TRUTH

MAPPINGS: Society/Theory/Space
A Guilford Series

Editors

GROUND TRUTH
The Social Implications of Geographic Information Systems
John Pickles, Editor

LAW, SPACE, AND THE GEOGRAPHIES OF POWER
Nicholas K. Blomley

WRITING WOMEN AND SPACE
Colonial and Postcolonial Geographies
Alison Blunt and Gillian Rose, Editors

TRAVEL, GENDER, AND IMPERIALISM
Mary Kingsley and West Africa
Alison Blunt

POSTMODERN CONTENTIONS
Epochs, Politics, Space
John Paul Jones III, Wolfgang Natter, and Theodore R. Schatzki, Editors

THE POWER OF MAPS
Denis Wood (with John Fels)

APPROACHING HUMAN GEOGRAPHY
An Introduction to Contemporary Theoretical Debates
Paul Cloke, Chris Philo, and David Sadler

GROUND TRUTH

THE SOCIAL IMPLICATIONS OF GEOGRAPHIC INFORMATION SYSTEMS

Edited by

JOHN PICKLES

THE GUILFORD PRESS
New York London

DEDICATED TO
THE MEMORIES OF

Brian Harley (1932–1991)
and
Michael Pickles (1954–1993)

A Division of Guilford Publications, Inc.
72 Spring Street, New York, NY 10012
Marketed and distributed outside North America by
Longman Group Limited.

Printed in the United States of America

This book is printed on acid-free paper.

Last digit is print number: 9 8 7 6 5 4 3 2 1

Library of Congress Cataloging-in-Publication Data

Ground truth : the social implications of geographic information systems / edited by John Pickles.
p. cm. — (Mappings)
Includes bibliographical references and index.
ISBN 0-89862-294-8.—ISBN 0-89862-295-6 (pbk.)
1. Geographic information systems. I. Pickles, John. II. Series.
G70.2.G76 1995
910′.285—dc20 94-31711
CIP

Contributors

MICHAEL R. CURRY Department of Geography, University of California at Los Angeles, California

MICHAEL GOODCHILD National Center for Geographic Information and Analysis, and the Department of Geography, University of California at Santa Barbara, California

JOHN GOSS Department of Geography, University of Hawai'i, Manoa

TREVOR M. HARRIS Department of Geology and Geography, West Virginia University, Morgantown, West Virginia

RONALD J. JOHNSTON Vice-Chancellor, University of Essex, Wivenhoe Park, Colchester, United Kingdom

RICHARD LEVIN Department of Sociology, University of Witwatersrand, Braamfontein, Johannesburg, South Africa

PATRICK H. MCHAFFIE Department of Geography, West Georgia College, Carrollton, Georgia

JOHN PICKLES Department of Geography and Committee on Social Theory, University of Kentucky, Lexington, Kentucky

SUSAN M. ROBERTS Department of Geography and Committee on Social Theory, University of Kentucky, Lexington, Kentucky

RICHARD H. SCHEIN Department of Geography and Committee on Social Theory, University of Kentucky, Lexington, Kentucky

PETER J. TAYLOR Department of Geography, University of Newcastle upon Tyne, Newcastle upon Tyne, United Kingdom

Timothy Warner Department of Geology and Geography, West Virginia University, Morgantown, West Virginia

Daniel Weiner Department of Geology and Geography, West Virginia University, Morgantown, West Virginia

Howard Veregin Department of Geography, Kent State University, Kent, Ohio

Preface

The gold rush is on. The hypermedia are happening. Whatever you call it, here come interactive graphics, text, video, all somehow user-chosen. But how will they tie together? If producers knew where all this was going, the rush would rival the advent of the Talkies. Meanwhile, each manufacturer says its gizmo will be the centerpiece of the hypermedia gold rush. (Nelson, 1992, p. 157)

A depiction is never just an illustration. It is the material representation, the apparently stabilised product of a process of work. And it is the site for the construction and depiction of social differences. To understand a visualisation is thus to inquire into its provenance and into the social work that it does. It is to note its principles of exclusion and inclusion, to detect the roles that it makes available, to understand the way in which they are distributed, and to decode the hierarchies and differences that it naturalizes. And it is also to analyze the ways in which authorship is constructed or concealed and the sense of audience is realised. (Fyfe & Law, 1988, p. 1)

The chapters in this book deal with the new technologies for gathering, analyzing, and mapping geographic data, and with the implications they have for our understanding of nature and social life. Over the past two decades the convergence of techniques for advanced computing and enhanced imaging has transformed the ways in which many of us think about and handle information. Together these technologies are now transforming our ways of worldmaking and the ways in which geographers and others think about and visualize the places, regions, environments, and peoples of the earth.

The development and deployment of sociogeographic data gathering, handling, and imaging techniques are part of a broader reconfiguration of the use of information in society. At the center of this reconfiguration is a renewed importance of the visual image. As visual representations are deployed in the new economy of information, spatial data handling and imaging techniques (of which geographic information systems [GIS] are a part) are being widely used in many areas of social life, business, and state management. GIS contributes to a (re)placing of the "visual" and the "spatial" at the center of social life through its role as an element in the restructuring of global, regional, and local geographies, the assertion of new disciplinary codes and practices, and the constitution of new images of earth and society.

Ground Truth: The Social Implications of Geographic Information Systems is, first, a book about the transformation of data handling and mapping capabilities that have emerged in the past two decades, and the impact they have had within the discipline of geography. Second, it is a book about the constellation of ideas, ideologies, and social practices that have emerged with the development of new forms of data handling and spatial representation. Third, it situates GIS as a tool and an approach to geographical information within wider transformations of capitalism in the late 20th century: as a tool to protect disciplinary power and access to funding; as a way of organizing more efficient systems of production; and as a reworking (and rewriting) of cultural codes—the creation of new visual imaginaries, new conceptions of earth, new modalities of commodity and consumer, and new visions of what constitutes market, territory, and empire. Our focus on GIS is thus also a reading (and a writing) of a broader history of the present, implicating issues of critical import to contemporary social transformation and the redefinition of what passes for democracy. The chapters approach these issues through analyses of technical change in specific contexts: in the discipline of geography, in the arena of production, in the use of advertising images, in the commodification of consumers, in the practice of war, and in the governance of territory.

Thus, beyond a political economy of technical change, *Ground Truth* also signifies an emerging economy of the virtual sign, of cybernetics and cyberspace. This emerging economy of information offers new possibilities for positive social action. As communications, information transfer, and imaging technologies (what I will refer to generically as telematics, informatics, and virtual reality or cyberspace) continue to colonize various domains of economic, social, and political life, they also permit the emergence of multiple communities of users with new capacities for action. The "information highway" penetrates the terrain of contemporary life, links formerly separate locations, gives rise to new imagined communities, and fosters new "spaces" for individual and collective identity. As an element of this round of modernization, GIS provides powerful data handling and mapping tools for these communities of action. Images of a whole earth, representations of relationships that transcend local, regional, or national identities, and new notions of community that transcend parochial conceptions of locality and place emerge as real possibilities through the intimate anonymity of electronic communication.

The Western trope of a public space in which people (usually "men") of good faith join in debate about their future, appropriated by industrial and urban forms of modernity as a mythic image of a democratic culture of debate and negotiation predicated on individual autonomy, private property, and state power, has more recently been further appropriated by the news and communication media through their claim to be the embodiment of the modern civic arena. This trope of public space is now being reappropriated by the electronic age as its wish image—the promise and possibility of "information." The putative openness of new electronic information media and the rhetoric of "voice," "openness," and "information"—the trope of reasoned, open, uncoerced discourse in a public place—is appropriated to the project of social development and private profit.

But, like all highways, the information highway requires points of access, capital investment, navigation skills, and spatial and cultural proximity for effective use. Like the automobile highway, the information highway fosters new rounds of creative destruction and differentiates among users and between users and nonusers. It brings regions of difference under a common logic and technology, and through differential access and use exacerbates old and creates new patterns of social and economic differentiation. While for some, information means the provision of alternatives and the satisfaction of choice (even if a "choice" signifies a socially constructed yet now naturalized whim of the wealthy consumer), for others this postindustrialism (and its attendant postmodern cultural forms) must still be seen in the context of a political economy of graft, monopolism, and uneven development.

Such processes of territorial colonization, globalization, and production of new scales of action contrast sharply with a technocultural ideology of enhanced autonomy and self-actualization, and severely complicates the assessment of the relationship between technological innovation and social change. Not only do data technologies treat all data and information within a universal logic and calculus, and not only do imaging technologies reach without break across socially and historically differentiated territories, but the tools themselves permit types of surveillant intervention that restructure everyday life itself. For some, this is a matter of market logic in which waves of competitive, leading-edge technologies are sufficient in themselves to drive the process of economic and social restructuring; the adoption of the technology by others is a sufficient (and necessary) reason for its adoption by us. Thus the dynamics of development and adoption are legitimized by an ideology of "progress" and an unproblematized belief in the importance of technical "advances" across such fields as science, medicine, administration, and logistics. New data handling and imaging capabilities are, in this way, fully naturalized as the next logical and necessary step in the advance of science and society, and the stimulus to new ways in which individuals and groups can overcome the barriers of distance and enhance their abilities to exercise control over society, space, and the earth.

But where technology is not seen as a social relation, it is fetishized and aestheticized, the contingent nature of technical outcomes is overlooked, and the struggles waged over the choice and application of any particular technology are ignored.

The chapters in this book, in various ways, all treat GIS as both technique and social relation, and each places GIS in a specific social context. In nearly all of these contexts the issues raised by spatial data handling and mapping techniques overlap with those raised by electronic imaging technologies more generally. Consequently, the chapters are also "readings" of the discourses, codings, and practices of contemporary geo-graphism in relation to the ways in which social life is being constituted by these more generalized informational technologies and virtual realities. In this sense, the possibilities, impacts, and limitations of new GIS technologies are located within a broader political economy of technical and cultural restructuring.

All the chapters are motivated in one way or another by a deep concern for the impacts of *unmediated technical practices* on the discipline of geography and other arenas of social life. All in some way address the impact of GIS on the reconfiguration of disciplinary knowledge, practices, and institutions, and the broader implications of these changes for the role of geographic research and teaching. Each raises important questions about the relations between science and society in a democratic society. The chapters explicate the nature of GIS as an emergent technology, as a way of doing research, *and* as a set of new social and institutional practices and relations. Some chapters focus on internal or relational aspects of GIS, while other chapters address broader metatheoretical and political issues emerging from the practices of users of GIS.

Chapter authors were given very general guidelines. No author or group of authors was in any way restricted with regard to content or approach. They were encouraged to reflect thoughtfully and critically on GIS within as wide a context as they wished, and to bring before the geographic and GIS community essays that will challenge thought and engage debate. Several chapters as a result, may, be unpalatable to contemporary users of GIS, and other chapters will be unsettling to those who look to this collection for a compendium of opposition viewpoints on the growth of GIS in the discipline or in its broader social applications.

The chapters include detailed case studies of the societal and disciplinary roles being played by technologies of surveillance. Some of the chapters focus on the disciplinary role of GIS and its effects on geographic practices. Others address more directly the ontological and epistemological status of the new electronic images. At least two of the chapters discuss the images produced and the ways in which they are being used. Some of the chapters address the widespread claim that GIS broadens the distribution of and access to information, reduces the burden of work on those who adopt its rigors and accept its benefits, and fosters democratic practice. Others address the political and social contexts within which such images are created and disseminated, while still others deal with the social and ethical implications of the new forms of representational object generated through electronic media.

While the surveillant and intrusive capabilities of new electronic systems and their appropriation to military and policing purposes must be addressed, we should at the same time account for the ways in which new ways of seeing the earth and new forms of resistance to state domination have also been made possible. Perhaps best known in this latter respect is the new image of earth as a whole made possible by NASA exploration, or the political space created for antisystemic strategies by electronic media during the Chinese democracy movement of spring 1989 and the Russian coup of 1990–1991 (Penley & Ross, 1991, p. viii). In these limited examples we see a glimpse of the decentralizing and empowering ability of the new information technologies to challenge the centralizing tendencies of state apparatuses or the private corporation. Vast changes in technological scope and scale force us to think in new ways about old approaches. With the emergence of postempiricist philosophies of science and poststructuralist theories of society, epistemologies based on representation as the mirror of nature have been fundamentally challenged, and with the development of virtual reality (in its many forms) even the nature of objects within a representational system arises as a question. An either/or logic has been rejected to some degree or other by all the authors who have contributed to the book. Each is in one way or another trying to come to grips with ways of responding to the scope and scale of these technologies and practices as they restructure social, economic, and political life.

In this context, the authors in this volume follow Constance Penley and Andrew Ross's (1991, p. xii) injunction that we must aim to "resist that tendency of fatalistic thought, and to include other kinds of stories that do not automatically fall in line with the tradition of left cultural despair and alarmism."

> Wary, on the one hand, of the disempowering habit of demonizing technology as a satanic mill of domination, and weary, on the other, of the postmodernist celebrations of the technological sublime, we selected contributors whose critical knowledge might help provide a realistic assessment of the politics—the dangers *and* the possibilities—that are currently at stake in those cultural practices touched by advanced technology. (Penley & Ross, 1991, p. xii)

Perhaps my central motivating theme for editing this collection is my belief that claims to truth (especially when articulated on the grounds either of competitive market necessity or of universal social benefit and more efficient and effective decision making) must be denaturalized, and what Denis Wood (1992) has called their "selective interestedness" must be unmasked. The discourses of "social progress," "technical advance," and "the postmodern condition" must, then, be situated within an analysis of the networks of power and systems of practice within which such discourses operate (and which they in turn partially constitute). If we are to make any claims about new technical capacities, uses, and limitations, we need to be clear about the different ways in which technologies affect specific groups and regions, reconfigure social relations, and increase the potential for the exploitation of some to the benefit of others. This is not to argue that new technologies emerge as social forces in a zero-sum context, but it is to recognize that the critical appropriation of the productive potential of any technology requires that its apparent "mere givenness" be denaturalized and its auratic hold be deconstructed.

This book was first planned by Brian Harley and myself in 1990. Brian died in December 1991 before the manuscripts were collected, but it has been my goal to edit and write in a way that is consonant with the intellectual project to which Brian contributed so much. More than anyone else in the discipline of geography, Brian Harley opened up the space for a critical theory of maps as instruments and inscriptions of power. It is to this power/knowledge tradition that this work contributes most directly.

I would like to thank a number of people who have helped in this project. Peter Wissoker of The Guilford Press has been a model of patience and encouragement during the various stages of delay through which the manuscript passed. The book has benefited from conversations with several colleagues besides those whose work it contains. Wolfgang Natter has been a kindred spirit in thinking the question of technology through the engagement of the works of critical theorists, hermeneuticists, and variously labeled poststructuralists. Others have, in one way or another, contributed to the ideas developed in the book: they are Bob Begg, Stanley Brunn, Jeremy Crampton, Deborah Dixon, Greg Elmes, Peter Gould, Derek Gregory, Robert Hanham, John Paul Jones, Doug Kellner, Michael Kennedy, Vickie Lawson, David Mark, Dimitrina Mikhova, Roger Miller, Gunnar Olsson, Tom Poiker, Mark Poster, Alan Pred, Mary-Beth Pudup, Dagmar Reichert, Eric Sheppard, Caedmon Staddon, Dick Ulack, Michael Watts, and Sam Weber. The University of Kentucky Department of Geography and the Committee on Social Theory have been supportive sites within which the question of technology and the engagement of critical theory, postructuralism, and the hypermodern can be thought.

REFERENCES

Fyfe, G., & Law, J. (1988). *Picturing power: Visual depiction and social relations.* New York: Routledge.

Harvey, D. (1989). *The condition of postmodernity: An inquiry into the origins of cultural change.* Oxford: Basil Blackwell.

Nelson, T. H. (1992). Virtual world without end: The story of Xanadu. In L. Jacobson (Ed.), *Cyberarts: Exploring art and technology* (pp. 157–169). San Francisco: Miller Freeman.

Payne, A. (1993). War in the age of intelligent machines. An interview with Manuel Delanda. *Public: Violence, 6,* 127–134.
Penley, C., & Ross, A. (Eds.). (1991). *Technoculture.* Minneapolis: University of Minnesota Press.
Wood, D. (1992). *The power of maps.* New York: Guilford Press.

Contents

CHAPTER 1

Representations in an Electronic Age

GEOGRAPHY, GIS, AND DEMOCRACY

John Pickles

The slow, uneven decline of these interlinked certainties, first in Western Europe, later elsewhere, under the impact of economic change, "discourses" (social and scientific), and the development of increasingly rapid communications, drove a harsh wedge between cosmology and history. No surprise then that the search was, so to speak, for a new way of linking fraternity, power and time meaningfully together. Nothing perhaps more precipitated this search, nor made it more fruitful, than print-capitalism, which made it possible for rapidly growing numbers of people to think about themselves, and to relate themselves to others, in profoundly new ways. (Anderson, 1983, p. 36)

DEFINING GIS

Defining geographic information systems (GIS) is not a straightforward matter. Even the use of the term "GIS" can be problematic. "GIS" refers to geographic information systems in the plural, yet "GIS" is often used as an acronym for a single system. Some writers choose to refer to "GIS systems," as systems of systems, while others have resorted to terms like "GISers" to refer to those with some strong commitment to GIS as a disciplinary enterprise.

GIS itself has a poorly developed archive and virtually no critical history of its own production, a fact recently emphasized by Coppock and Rhind (1991):

> A variety of information indicates that the field of GIS has expanded rapidly in recent years. . . . From where did all this business and the resulting jobs arise? Unhappily, we scarcely know. GIS is a field in which history is little more than anecdotal. To rectify this, a search through the archives of government departments and agencies would certainly help. As yet, however, few organizations have given any thought to formalizing the history of their involvement in GIS and at least one major player (Ordnance Survey . . .) has refused to let its detailed records be examined by external researchers. (p. 21)

Moreover, the definition of GIS varies depending upon who is giving it, and whatever definition we do give it is likely to change rapidly as digital spatial data and computer graphics spread rapidly into engineering, medical, earth science, design, planning, and other fields.

Central to each of these possible definitions is some relational system of spatial information handling and representation. GIS is a special case of information systems in general, in which information is derived from the interpretation of data "which are symbolic representations of features" (Maguire, Goodchild, & Rhind, 1991, p. 10). The designation GIS is also "frequently applied to geographically oriented computer technology, integrated systems used in substantive applications and, more recently, a new discipline" (Maguire et al., 1991, p. 9). Here GIS refers to the integration and use of "computer-aided design, computer cartography, database management and remote sensing information systems" (Maguire et al., 1991, p. 12), to the mapping of information using digital technology (Newell & Theriault, 1990, p. 42), or to any kind of automated geographic data processing (K. C. Clarke, 1986).

These competing definitions are reflected in differences in interpretation of the central principles of GIS. Maguire et al. (1991, pp. 13–14) suggests three such undergirding principles: that GIS focuses on the cartographic display of complex information; that GIS is a sophisticated database system; and that GIS is a set of procedures and tools for fostering spatial analysis. However, "the recent origin and rapid rate of progress has not been conducive to the analysis and definition of GIS" (Maguire et al., 1991, p. 9). Part of the reason for this is said to be the commercial nature of the commodity, which leads sellers and developers to produce a "great deal of hyperbole and rhetoric" and to offer conflicting advice and information. Part of the reason has to do with the ways in which GIS has developed within different disciplines and research contexts (in agriculture, botany, computing, business, photogrammetry, geology, zoology, surveying, engineering, and geography), for each of these fields puts its own peculiar stamp on the claims it makes for GIS. Thus, "together these factors [those mentioned above plus others] have conspired to obfuscate an issue which has never really been satisfactorily discussed or analysed in any detail" (Maguire et al., 1991, p. 10).

Two of the central defining characteristics of all geographic information systems are the role of digital electronic data and the production of electronic spatial representations of those data: GIS is a product of computers in particular and of electronic information technology more generally. When we turn to such electronic media, to what objects do we turn? Do we turn to the objects themselves, the artificial neural networks that facilitate data entry, capture, and reproduction? Do we turn to the speed with which new devices allow us to operate and communicate? Do we turn to new forms of representation that are allowed and disseminated by new devices and apparatuses? Or do we turn to new practices that are not intrinsic to the new media, but are permitted and facilitated by them: technologies of the body, of the social, of the economy, by which bureaucratic, business, or military functions (and others) can be extended effectively across new territories with effects that previous technologies did not permit? I think we must admit that in our enthusiasm and confusion, we refer to all of these at once; that is, that (like geographic information systems themselves) the new electronic media produce multiple overlapping effects with which it is analytically and politically difficult to deal.

GIS thus operates at several levels and the term "GIS" refers to several distinct types of object: a research community that transcends disciplinary boundaries; an approach to geographical inquiry and spatial data handling; a series of technologies for collecting, manipulating, and representing spatial information; a way of thinking about spatial data; a commodified object that has monetary potential and value; and a technical tool that has strategic value. Academic developers and users of GIS have tended to focus primarily on the technical and organizational issues raised by the use of electronic information and imaging. But because of the high cost of its development and use, GIS has emerged above all as a tool and product that changes the way certain groups and organizations operate. That is, it is a technology (like all technology to one degree or another) closely tied to the concrete material and ideological needs and interests of certain groups. As such, it is an important element in changing social relations in market economies; in producing new demands, commodities, and forms of domination in the workplace; in developing new systems of counting and recording populations; in defining, delimiting, and mapping space and nature; and in providing new tools and techniques for waging war. In each of these domains GIS is part of a contemporary network of knowledge, ideology, and practice that defines, inscribes, and represents environmental and social patterns within a broader economy of signification that calls forth new ways of thinking, acting, and writing.

Despite this ambiguity and the absence of coherent definition, the development and adoption of these new information and imaging technologies is increasingly being referred to as a revolution—almost Maoist in form—in which new technologies succeed each other in ever shorter periods of time, and as a result of which speed of interactions is increased, unit costs are reduced, and new methods are applied to old (and new) problems. Already "projections for the 1990s indicate that the GIS field will grow by as much as 35 to 40 percent, based on projected sales of GIS-related hardware, software, and services" (Huxhold, 1991, p. 12), and boosters are already proclaiming the emergence of a new profession, the GIS profession. But in this emerging profession the question "Is a geography degree the ticket to GIS success?" (Huxhold, 1991, p. 20) is posed alongside the questions "What is the GIS profession, what does it take to be a part of it, and what does it pay?" (Huxhold, 1991, p. 22). As Maguire et al. (1991, p. 17) point out, "GIS are clearly big news" and "it is not fanciful to suggest that by the end of the century GIS will be used every day by everyone in the developed world for routine operations." What is not yet clear is, what forms of change and what kinds of distortion will result from these patterns of adoption if the discipline strives to retain a central role in this emerging "profession"?

GIS is a set of tools, technologies, approaches, and ideas that are vitally embedded in broader transformations of science, society, and culture. These contexts are wide-ranging and as yet little studied in the literature surrounding new mapping and analytical technologies, including GIS. But the questions are gradually being raised in the broader contexts of mapping generally (Harley, 1988a, 1988b, 1989, 1990; Pickles, 1991, 1992b; Smith, 1992; Hall, 1993; Wood, 1992), virtual reality (Rheingold, 1992; Rosenthal, 1992; Wooley, 1992), and cyberspace (Benedikt, 1991; Crary & Kwinter, 1992; Penley & Ross, 1991).

In this book, while addressing questions of immediate relevance to GIS and geography, each of us is also struggling with the complex of technologies and practices with which GIS is associated (electronic media, cyberspace, virtual reality, new disciplinary practices) and the contexts within which they have emerged (internationalization, globalization, commodification of information, market penetration). But this complex of technologies has been poorly defined within a language and framework that weakly reflects its impacts on issues such as individual autonomy, privacy, access, systems of governance, marketing strategies, and military tactics. We are, that is, entering a potential new phase of ways of *worldmaking* for which we desperately need new ways of *wordmaking* (see, e.g., Luke, 1993; Olsson, 1992; Pred & Watts, 1992; Ronnell, 1989).

The task of definition in this sense is too important to be left only to the experts. As GIS enters into the fields of public policy, regional planning, business, the military, and private lives, its effects are wide-ranging and the issues its application raises are important. However, with notable exceptions (such as the resignation from Congress in 1987 of Congressman George Brown in protest against the almost exclusive use of satellite technology for reconnaissance purposes and the fact that a ban existed on public discussion of the issue (Barry, 1992, p. 571), the development and application of GIS have rarely been treated as having serious political and social implications. Moreover, for the most part, GIS users themselves have failed to address these wider contexts of practice and meaning within which their own activities are located.

Thus, it would be wrong to see the new informational databases as merely more efficient counting machines. The effort that has gone into their development and production signals a broader restructuring of the economy of information within which they are put to use. In this sense, the recent thrust to develop and diffuse institutional and professional foundations for these new technologies and tools also signals the wider instanciation of this new economy of information in society—an economy of accounting, recording, archiving, overlaying, cross-referencing, and mapping information. If we are to seriously engage the disciplinary and social roles played by GIS, it must be contextualized within these broader (and in some ways more dynamic and problematic) developments in representational technology generally.

THE TECHNOLOGY AND ITS POSSIBILITIES

Insofar as it has enlarged our vision of how data and information can be linked in new and interesting ways, GIS has brought about far-reaching and significant changes within scientific research, public and private agencies, and the disciplinary structure of geography. Like the market-oriented communications and information systems that are currently gaining ground within liberal democracies, GIS technologies and programs of research and teaching are being sold to the geographic profession and to the broader public "on the promise that it will enlarge people's choices and increase control over their lives, that is, that it will be both liberating and empowering. This emerging order is the product of two major processes: technological innovation and convergence, and 'privatization' " (Murdock & Golding, 1989, p. 180).

In *Mapping the Next Millennium: How Computer-Driven Cartography Is Revolutionizing the Face of Science,* Hall (1993) has recently pointed to this issue through an analysis of new mapping technologies across a wide range of the sciences. His analysis deserves more detailed consideration, as a beginning point for our analysis of the economy of mapping and representation. Hall (1993, p. 8) suggests that, fueled by new facts and new systems of instrumentation, we are in the middle of "arguably the greatest explosion in mapping, and perhaps the greatest reconsideration of 'space' (in every sense of that word)" since the times of Babylon, and that this redefinition requires a rethinking and broadening of our conceptions of maps and mapping:

> With stunningly precise new instruments of measurement developed over the last half century and with the tremendous graphic powers provided by computers over the last two decades, everyone from archaeologists to zoologists has been able to discover, explore, chart, and visualize physical domains so remote and fantastic that the effort involves nothing less than the reinvention of the idiom of geography. (Hall, 1993, pp. 4–5)

Part of this redefinition involves the incorporation of technically precise methods. These methods encourage concern for the "perfect GIS" in which the base map would be accurate and geodetically correct, data would be available in compatible forms and formats, the GIS would be maintained and current, and all sorts of information—from state boundaries to 3-D models of grocery stores—would be included

(Abler, 1993, p. 132; Keating, 1992, p. 32). The integration of the technology of accurate location—in this case an integrated GIS/GPS (Global Positioning System)—would permit an improved geography to be developed, a three-dimentional representation (a geography in depth) accurately pegged to the material world around us. The modeling of human and environmental interactions in this new global geography and "global geography machine" will be possible if GIS is tied to GPS, and mapping is reduced to the accurate representation of the materiality of the earth and to the accurate determination of position (Abler, 1993).[1]

Abler's (1993) discussion of GIS/GPS exemplifies the current concern for data and accuracy at the core of data collection and management techniques. But such claims also have the effect of directing the attention of geographers away from the broader field of spatial representations with which the new global imaging systems are already being merged, specifically the world of virtual, not "real," realities. Ted Nelson (1992, p. 158), in contrast, has argued strongly that "our world becomes increasingly virtual, as its appearance departs more and more from depending on the structure of physical reality." And this notion, it seems to me, captures more effectively the spatiality of GIS—a virtual space of data manipulation and representation whose nominal tie to the earth (through GPS and other measuring devices) is infinitely manipulatable and malleable. The Newtonian world of Abler's GPS/GIS fusion, although technically necessary, seems a skeletal form compared to the virtually chaotic, complex worlds of fractal space, hypertext, and GIS.

In this new world of cartographic experimentation, technology generates its own appropriate concepts of scale: "My approach is to try to get people to *drop* human scale completely. And when they think of something, they go into *that* scale. If you're going to think of galaxies, you've got to be galaxy-like. . . . If you don't expand yourself to that scale, I think it's hopeless" (interview with Alan Dressler, Aug. 28, 1989, in Hall, 1992, p. 5). Digital spatial data and GIS permit the infinite manipulation of data layers, the construction of an infinite sequencing of new views on the data landscape, new angles of view, multiple overlays, and correlations of spatial data landscapes. Space and data have become fully manipulatable in this virtual environment.

One of the more popular recent expositions of this wider perspective is David Gelernter's (1992) *Mirror Worlds or the Day Software Puts the Universe in a Shoebox . . . How It Will Happen and What*

It Will Mean. *Mirror Worlds* is a popularized introduction to the goals and visions underpinning the development of virtual worlds, but it provides a useful point from which to view the epistemological assumptions and social claims within this broader field of virtual spatial realities. *Mirror Worlds* "describes an event that will happen someday soon. You will look into a computer screen and see reality. Some part of your world—the town you live in, the company you work for, your school system, the city hospital—will hang there in sharp color image, abstract but recognizable, moving subtly in a thousand places" (Gelernter, 1992, p. 1). The mirror world of virtual reality and spatial images is a "true-to-life mirror image trapped inside a computer—where you can see and grasp it whole" (p. 3). These images "engulf some chunk of reality" (p. 6) and the mirror world "reflects the real one" (p. 6). "Fundamentally these programs are intended to help you *comprehend* the powerful, super-techno-glossy, dangerously complicated and basically indifferent man-made environments that enmesh you, and that control you to the extent that you don't control them" (p. 6).

How is this to happen? How will the "place" of the mirror world permit one to enter, stroll around, and retrieve archival and live-medium information?

> The picture you see on your display represents a real physical layout. In a City Mirror World, you see a city map of some kind. Lots of information is superimposed on the map, using words, numbers, colors, dials—the resulting display is dense with data; you are tracking thousands of different values simultaneously. You can see traffic density on the streets, delays at the airport, the physical condition of the bridges, the status of markets, the condition of the city's finances, the current agenda at city hall and the board of education, crime conditions in the park, air quality, average bulk cauliflower prices and a huge list of others. . . .
>
> This high-level view would represent—if you could achieve it at all—the ultimate and only goal of the *hardware* city model. In the software version, it's merely a starting point. You can dive deeper and explore. Pilot your mouse over to some interesting point and turn the *altitude* knob. Now you are inside a school, courthouse, hospital or City Hall. You see a picture like the one at the top level, but here it's all focussed on this *one* sub-world, so you can find out what's really going on down here. Meet and chat (electronically) with the local inhabitants, or other Mirror World browsers. You'd like to be informed whenever the zoning board turns its attention to Piffel Street? Whenever the school board finalizes a budget? Leave a software agent behind. (Gelernter, 1992, pp. 16–17)

The elaboration of new virtual worlds and spatial images extends our own world and our thinking about that world in remarkable ways. In this context, GIS is merely one part of a larger tradition of digital data handling and spatial representation. Part of this wider tradition includes multimedia and hypertext. Mark Poster (1990) has perhaps provided the most thorough theorization of the new revolution in visualization brought about by new electronic information systems, but it is in the work of G. P. Landow (1992) that poststructuralist ideas are brought directly to bear on an interpretation of multimedia and hypertext. For Landow, critical social theory promises a way of theorizing hypertext, and hypertext embodies and tests theories of textuality, narrative, margins, intertextuality, and the roles and functions of readers and writers. In Roland Barthes's term, hypertext produces "writerly" texts that do not dominate the reader and insist on particular readings, but instead engage the reader as an "author" and insist upon the openness and intertextuality of the text—that is, its openness to other texts and readings.

> When designers of computer software examine the pages of *Glas* or *Of Grammatology,* they encounter a digitalized, hypertextual Derrida; and when literary theorists examine *Literary Machines,* they encounter a deconstructionist or poststructuralist Nelson. These shocks of recognition can occur because over the past several decades literary theory and computer hypertext, apparently unconnected areas of inquiry, have increasingly converged. (Landow, 1992, p. 2)

Such information-handling and imaging technologies place the visual at the very center of the known, and raise important questions about the nature of the image. As Walter Benjamin (1968, pp. 233–234) has shown us, there are important differences between the image produced by the camera and the image produced by the painter: "The painter maintains in his work a natural distance from reality, the cameraman penetrates deeply into its web. There is a tremendous difference between the pictures they obtain. That of the painter is a total one, that of the cameraman consists of multiple fragments which are assembled under a new law." With the emergence of spatial digital data, computer graphic representation, and virtual reality, the law has changed again. The principle of intertextuality common to both hypertext and GIS directs our attention to the multiple fragments, multiple views, and layers that are assembled under the new laws of ordering and reordering made possible by the microprocessor. Some have even suggested that virtual reality signals the end of photography as evidence for anything, or that virtual representations will produce illusions that will be so powerful it will not be possible to tell what is real and what is not real.

Such claims are deeply disturbing and at root problematic. Nonetheless, GIS and informatics do open virtual spaces for "real" social interaction, new communities of dialogue, and new interactive settings for which we currently have only poor language and no architecture. The electronic airwaves are, in this view, interpreted to be foundational for the reemergence of a civic culture, a community of dialogue, and a global village. They are also the potential source of new powers for marginalized groups to whom traditional media have been inaccessible. In this view, the electronic airwaves and systems of informatics provide a potential source of counterhegemonic social action, and GIS—as a specific form of data handling and imaging—offers a diverse array of practical possibilities. In both cases (the resurgence of civic culture and the potential for counterhegemonic action) informatics are seen as a potential liberator of socially and politically marginalized groups, and thus as a source of democratizing power for these newly networked groups. If information is power in this sense, and if community is built through dialogue, then informatics permits both to emerge for those who would otherwise have no voice and no space for collective action.

Uses of communication systems for politically progressive purposes and for the defense of speech against totalitarianism have recently taken on a character of mythic proportions, as users extol the progressive uses of the fax machine by students in China during the spring 1989 Democracy Movement, or the use of e-mail by those opposed to the coup against Gorbachev to maintain contact with each other and the outside world (Penley & Ross, 1991, p. viii). Like Pancho Villa, who captured the trains and used them to attack government troops and gain access to the very heart of the cities during the Mexican Revolution, new informatic democrats and revolutionaries are eulogized as examples of progressive power, and as counterexamples to the more widespread business, state, and military uses of the technology.

Even though the funding for research and development of the hardware and software used in GIS and other imaging systems has come primarily from business, state, and military sources, advocates of the progressive potential of information and imaging technologies argue that access is hard to deny, networks are quite difficult to control, information is readily accessible and used by individuals and groups with limited budgets and expertise, and the ability to use the technology in depth permits groups like environmental organizations to counter claims by polluters about their environmental impacts, by developers about likely local effects of runoff and groundwater, and so on. In this view, GIS enables communities to make better decisions by providing access to more and better information. It provides more powerful tools for local planning agencies; it offers exciting possibilities for data coordination, access, and exchange; and it permits more efficient allocation of resources, and a more open rational decision-making process.[2]

EPISTEMOLOGICAL INERTIA AND THE NEW IMPERIAL GEOGRAPHY

> *It is not fanciful to suggest that by the end of the century GIS will be used every day by everyone in the developed world for routine operations. (Maguire et al., 1991, p. 15)*

The discussion in the previous section detailed elements of a socially and critically engaged role for GIS. It built largely on the claims of proponents of GIS and informatics generally about the possibilities inherent in the technology and/or the use of the technology. Not all geographers have been happy to accept some or any of these claims made by GIS users, and in this section I ask whether such claims are sustainable given questions about the underlying assumptions, ontology, and politics of GIS and its use. If we determine that these claims are not sustainable, do we know what a critical use of GIS and imaging technology would require? If the claims of GIS users can be sustained, what should be the relation between GIS and critical science, and what effects would such a situation create for critical studies in geography, particularly in those branches of the discipline (and of many other disciplines) where "the positivist assumptions embraced by GIS have long since been jettisoned" (Lake, 1993, p. 404)? Can we transform GIS and other imaging technologies to make them compatible with the premises and commitments of critical science? Or can we rethink our understanding of the new information and imaging systems in ways that will allow their productive potential to be deployed in new ways?

The self-understanding of GIS itself can be readily observed in the two-volume *Geographical Information Systems* (Maguire et al., 1991).

This publication—the most thorough synthesis and analysis of GIS to date—is a vast compendium of the history, principles, tools, and methods of GIS: an encyclopedia and handbook for GIS, a marker of the state of the field at the present time, and a workbook for teachers and students to deepen their awareness of the field. Maguire et al.'s text provides the first solid support for the claim that GIS is entering into a new phase and approaching the possibility of creating a separate discipline—a claim made by the editors of the book (particularly in the Introduction) and boosters within the commercial sphere (such as *GPS World*). Indeed, the work reflects the emergence of strong research agendas within GIS over the past two decades, as well as the close integration of academic, public, and commercial developers and users of GIS.

The book is also a marker of another aspect of disciplinary history. It is perhaps the most comprehensive statement to date of a branch of geographic scholarship that has systematically pursued a vision of the geographic, with an epistemology and belief in method quite different from that pursued by large parts of the rest of the discipline. While within geography and the social sciences generally the period from the 1960s to the 1990s saw the emergence of new paths and principles, which—from the critique of objectivist science in the 1970s to the impacts of postmodernism and poststructuralism in the 1990s—have transformed the central questions and approaches of the discipline, the epistemology and method that underpins GIS emerged in the 1960s under the auspices of positivist and empiricist versions of science and reemerged as a result of the collaboration between, and a revitalization of, spatial analysis, cybernetics, and computer developments of the 1970s.

Taylor's (1990, pp. 211–212) trenchant critique of GIS as the new imperialist geography suggests that GIS has emerged as a two-part strategy on the part of unreconstructed "quantifiers" who have "bypassed" the critiques levied against the empiricism of spatial analysis, and at the same time have captured the rhetorical high ground of a progressivist modernism (or naive postindustrialism) by readily accepting the switch from knowledge to information:

> Knowledge is about ideas, about putting ideas together into integrated systems of thought we call disciplines. Information is about facts, about separating out a particular feature of a situation and recording it as an autonomous observation. . . . The positivist's revenge has been to retreat to information and leave their knowledge

> problems—and their opponents—stranded on a foreign shore. But the result has been a return of the very worst sort of positivism, a most naive empiricism. (Taylor, 1990, pp. 211–212)

In this (re)turn the geographical is defined as the study of anything that is spatial[3]:

> GIS is a technological package that can treat any systematic collection of facts that are individually identified spatially. These facts may be medical statistics, remote-sensing images, crime files, land-use data, population registers or whatever. In terms of the package, spatial patterns can be produced irrespective of what the information is about. . . . Such quantifiers can produce a maverick geography dealing with crime one week, bronchitis the next, and so on. (Taylor, 1990, p. 212)

The colonizing aspirations of such claims and such an approach are—as Taylor points out—transparent. Many GIS users undoubtedly see these claims as exaggerated at best and false at worst, or, as Openshaw (1991) has argued, they represent reductionist assertions and derogatory and confrontational language, "knockabout stuff" that emerges from a reactionary desire to protect a particular system of order and power. Thus, for Openshaw, the crisis to which Taylor points is redefined as "contrived" and should be replaced by a notion of "creative tensions" between at times complementary, at times competing, but equally productive, intellectual projects. Oppenshaw (1991) describes the possibilities of GIS:

> A geographer of the impending new order may well be able to analyse river networks on Mars on Monday, study cancer in Bristol on Tuesday, map the underclass of London on Wednesday, analyse groundwater flow in the Amazon basin on Friday. What of it? Indeed, this is only the beginning. (p. 624)

According to Openshaw (1991. p. 626), this new-order geography needs GIS in order to "put the pieces of geography back together again to form a coherent scientific discipline." He continues:

> It would appear then that GIS can provide an information system domain within which virtually all of geography can be performed. GIS would emphasise an holistic view of geography that is broad enough to encompass nearly all geographers and all of geography. At the same time it would offer a means of creating a new scientific look to geography, and confer upon the subject a degree of currency and relevancy that has, arguably, been missing. (p. 626)

This imperialist, reductionist, and technicist view of GIS (and geography) is further illustrated in Martin's (1991) *Geographic Information Systems and Their Socioeconomic Applications.*[4] The book is important because it is one of the few to explicitly address the role of GIS in socioeconomic applications, and because it does so explicitly from an understanding of GIS as a spatial analytic and applied science. Martin begins with a discussion of the absence of any clear theoretical structure guiding the developments of GIS, and argues that "to an outsider GIS research appears as a mass of relatively uncoordinated material with no core theory or organizing principles" (p. 44). In order to overcome this absence of theoretical work, Martin defines "spatial data" and "geographic data" by using the definitions developed for spatial analysis in Abler, Adams, and Gould's 1971 volume *Spatial Organization of Society:* "spatial data" is a general term used to refer to measurements that relate objects existing in space at any scale, and "geographical" data is a term used to refer to data relating to objects in the range from architectural up to global scales. Geography is the analysis of objects and patterns in space (not, coincidentally, exactly the subject matter of GIS), or as Martin (1991, p. 45) says, quoting Abler et al. (1971), "Almost any substantive problem a geographer tackles can be fruitfully considered as a problem of describing accurately and explaining satisfactorily the spatial structure of a distribution." Martin builds upon and revitalizes—as the basis for a theory of GIS—these theoretical concepts of space and geography, but he does so in such a way that they remain disengaged from any consideration of the broader discussions and theoretical debates that have transformed the discipline in the past 20 years, and even from those efforts that sought to rethink concepts of space within spatial analysis (see, e.g., Gatrell, 1983). These debates about the nature of space, spatial objects, and what constitutes geographical objects are ignored here in the interest of reconfiguring "theories of GIS" in terms of purely abstract spatial objects and the relations between them.

Representing the natural and social world is, in this view, only a technical problem: analog models or real maps of the real world provide a model that is "an 'accurate' representation of the world and . . . embodies the spatial relationships necessary for the solution of any particular problem" (Martin, 1991, p. 48). In an attempt to ground the special category of representation that deals with socioeconomic phenomena, Martin turns to the theory of maps and lays out the traditional cartographic model of the relationship between the real world and the map as a model of the real world. In this model, "the cartographer's task is to devise the very best approximation to an 'ideal' transformation involving the minimum of information loss" (Martin, 1991, p. 54).[5] In struggling with the problem of representation, Martin resolves the matter by recourse to a traditional positivist interpretation of maps, in which the relationship between reality and image is an unproblematic one of representation, and error is merely a result of lack of technical skill or unintentional distortion (for a wider discussion of this model of cartography, see Pickles, 1992b). The result is a book that represents (in spite of its clarity and careful definition, among other strengths) the kind of theory of GIS that leads some geographers to argue that GIS can be seen as a form of unreconstructed (or only partially reconstructed) spatial analysis operating with assumptions derived from a positivist tradition. Its concepts and epistemology of space, objects, and reality are taken directly from the spatial analytic tradition. Martin argues for a straightforward observer epistemology and a view of theory as definition. To this spatial analytic tradition are added computers, power, and flexibility. But many of the old problems remain. This is reflected in Martin's conception of the socioeconomic applications for which GIS is particularly well suited:

> Unlike a road intersection or a mountain summit, we are rarely able to define the location of an individual simply by giving their map reference. This has far-reaching implications: socioeconomic phenomena such as ill health, affluence and political opinion undoubtedly vary between different localities, but we cannot precisely define the locations of the individuals which make up the chronically sick, the affluent or the *politically militant.* If GIS are to be used to store and manipulate such data, it is crucial that much care is given to ensuring that the data models used are an acceptable reflection of the real world phenomena. (p. 5; emphasis added)

> GIS may become invaluable to the efficient functioning of organizations. (p. 40)

> The growth of these data and their use in relation to socioeconomic phenomena has become known as "geodemography." Many organizations, including health authorities, retailers, and direct-mail agencies have become very interested both in the description of geographic locations in terms of their socioeconomic characteristics, and the identification of localities containing people of specific socioeconomic profiles (e.g., poor health, high disposable income, etc.). (p. 41)

The prospect of socioeconomic applications of GIS permitting efficiently functioning organizations such as insurance companies to develop "geodemographical" insurance rate schedules based on the identification of zones and localities of high risk, the targeting of civil rights groups (the "politically militant") for particular police or vigilante attention, or the extension of direct-mail solicitation to exact market targeting based on recorded purchasing and general expenditure records (already a reality, of course), seem actually to be applauded in Martin's eagerness to "sell" to the reader the potential socioeconomic applications of GIS. Martin takes as unproblematic what has become naturalized practice within the GIS community. Thus, for example, it is difficult to distinguish between the unabashedly boosterist claims of the academic Martin from the unabashedly boosterist claims of the business authors in the October 1991 issue of *GIS World,* which lauded the role of "GIS in Business" (Francica, 1991; Maffini, 1991) with fluffy articles dealing with socioeconomic applications ranging from real estate, energy delivery, agribusiness, tourism, and communications (Dangermond, 1991), to the insurance industry (Runnels, 1991), to retailing (Moloney & Dellavedova, 1991), to market area analysis for car dealerships (Clark, 1991), to fleet management (Barry, 1991), to delivery services (Heivly, 1991), to direct marketing (Moncla & McConnell, 1991; Cooke & Plumer, 1991), to telecommunications (Gusso & Lasala, 1991), to fast food location strategies (Kirchner & Thomas, 1991).

In writings concerned with the social and economic applications of GIS the absences and silences are particularly instructive. Whole domains of praxis within which GIS might make some contribution are elided, and Martin (and much GIS) remains silent about them. Instead, the gaze of the strategic planner, the commercial manager, or the military strategist is presented as an appropriate application—this is the kind of technocratic myopia that led Gunnar Olsson from 1972 on to charge that spatial analysis was an inherently conservative form of analysis (Olsson, 1972, 1974). In this myopic vision, there is rarely room for the insurgent GIS, or for GIS socioeconomic applications other than those that permit us to gain greater levels of clarity and control over the social and economic domain.

Moreover, Martin (as do many others writing about GIS) fails to ask any questions about the current trajectory of GIS research and practice. No attention is given to the question of the scale and cost of technology and its relation to specific types of socioeconomic application.

No reference is made to the growing amount of Third World literature on pc-based GIS for local action groups, or the use of computerized databases to monitor and control polluting state enterprises in centrally planned economies, or to the flourishing of disparate efforts by progressive GIS users to develop networks of local, small-scale systems to provide information that challenges the very corporate and statist interests that Martin seems to see as the major users of "socioeconomic applications."

Martin's book typifies a strong thread in the emergence of GIS as a disciplinary discourse and social practice. The book presents only an implicit and indirect picture of the representational economy emerging within contemporary GIS and its relationship with an economy of control. In this economy, socioeconomic applications are aimed at organizational efficiency and control of geographic territory (be it the jurisdiction of a health, police, or military authority, or the market area of retailers or direct marketing agencies). Implicit is the view that if data and technology availability permit the manipulation of spatial data for particular ends, then the ends themselves are justified (or of no concern to the geographer). Missing is any analysis of the ethical and political questions that emerge as GIS institutions and practices are extended into socioeconomic domains. That Martin's book is intended to be a text on the socioeconomic applications of GIS for students and GIS users (like the bulk of texts dealing with GIS), but lacks any treatment of ethical, economic, and political issues, raises serious questions about the possibility for the emergence of critically and socially responsible behavior within this particular episteme and its associated practices.

Like Martin, GIS authors more generally have grounded their analyses in terms of value-neutral observation, science as the mirror of reality, and theory as the product of data collection and testing, and have not chosen to engage the disciplinary and social theoretic debates of the past two decades that address the intellectual, social, political, and technological impacts of this form of instrumental action. In speaking about planning and applied geography, Robert Lake (1993) explicitly ties the development of GIS to this "resurrection" of a rational model of planning and a positivist epistemology:

> The unrelenting embrace of the rational model by planning and applied geography is not adequately described merely in terms of the tenacity and inertia of convenient and familiar practices. The rational model has been actively resurrected and rehabilitated by the ascend-

> ance of Geographic Information Systems to a position near to or at the core of both planning and geography. (p. 404)

Lake's claim that positivist epistemologies have been resurrected and rehabilitated can, I think, be sharpened even further. While it is certainly the case that many critical theorists in geography see in GIS a rehabilitation of positivist epistemology, from a different perspective it is clear that positivism was never forsworn, nor was the critique of positivism seriously engaged by GIS scholars. Empiricist and positivist assumptions continued to ground spatial analytic work throughout the 1970s and 1980s, and they were not seriously challenged either by the turn to behaviorist and behavioral geographies in the 1970s or by the turn away from analytic traditions toward humanism and subjectivist epistemologies in the 1980s (see Pickles, 1985, 1986). Instead of a thorough engagement with the epistemological debates that emerged in the social sciences in the 1970s and 1980s, or with the linguistic turn in the 1980s, geographers working in the field of GIS merely sidestepped into the research tradition of artificial intelligence, whose heritage of cognitive simulation and semantic information processing provided a compatible intellectual and professional home for work on automated cartography and GIS software.[6]

From this perspective, it becomes clearer why Lake's (1993, p. 405) review of the literature finds few publications on the part of GIS proponents that consider the epistemological, political, and ethical critiques of positivism, or any serious engagement with what he terms the "fundamental disjuncture growing at the core of the disciplines." That disjuncture is perhaps even more severe than Lake suggests. Such attempts to ground geographic research methods (and for Openshaw, 1991, to ground geography as such) in automated data handling seem to many GIS critics strangely distanced from contemporary theoretical debates about geographical method, epistemology, and ontology. In the 1980s, human geography developed strong critiques of the reductionist ontology of spatialism and turned to questions of contextual knowledge; contingency and necessity; society, space, and nature; the (social/political/gendered) construction of space; and the production of scale—each of which in various ways problematized aspects of Cartesian science and the ontology of spatial analysis. These approaches questioned the overemphasis on pattern, challenged geographers to rethink the meaning of space, problematized the dominance of natural science methodology in the study of social phenomena, and raised questions about the underlying ontology of objects, location, and application on which spatial analysis was predicated. Lake's disjuncture is, in this sense, grounded in a fundamental epistemological divide between positivists and postpositivists, between Kantians and those who have heeded the extended debates generated by the linguistic turn and the interpretive turn, and—as Derrida has coined it—between those whose feet are firmly planted in the soil of logocentrism and the metaphysics of presence, and those who have taken on board the implications of the critique of logocentrism and ontotheology. The former claim to stand on the critical tradition of empirical science, while the latter call for a critical theory that engages the logic of limits, the marginal, and the liminal; that overturns many of the epistemological assumptions on which such a science can be built, and that locates a new understanding of power at the heart of claims to knowledge (Derrida, 1976; Norris, 1987).

> Perhaps more troubling is the likelihood that consideration of these issues will be even further obscured by the popular momentum, technological complexity, and sheer scale of financial investment represented by the ascendancy of GIS. Once that investment is made, the focus is more likely to turn to expanding applications than to reconsidering philosophical foundations. (Lake, 1993, p. 405)

Brian Harley (1990, p. 8) had made exactly this point in his review of contemporary computer cartography: "We can glimpse here the unconscious process of myth-making, through which the invention of a progressivist positivist past is used to justify a progressive positivist present."

FROM SABER-RATTLING TO ENGAGEMENT

> *The essence of technology is by no means anything technological. Thus we shall never experience our relationship to the essence of technology so long as we merely conceive and push forward the technological, put up with it, or evade it. Everywhere we remain unfree and chained to technology, whether we passionately affirm or deny it. But we are delivered over to it in the worst possible way when we regard it as something neutral; for this conception of it, to which today we particularly like to do homage, makes us utterly blind to the essence of technology. (Heidegger, 1977, p. 4)*

GIS technology has, from its early days, been big business. Currently it is huge business, and the scale and scope of this business is not hidden in the marketplace. The power of data handling now means that GIS and related data handling and imaging systems have become central elements in demographic and infrastructural accounting systems; international, national, and regional monitoring and management projects; business organization; design activities; and military weapons and strategic planning. Since electronic information technologies provide more information and faster access across broader spans of space, they are presumed to be technologies that are liberating. Such a mythos of public benefit accruing from the *ability* to gain access to new and broader forms of data, and to represent this data spatially in a wide array of images, has been instrumental in the adoption of the new telematics within universities, planning agencies, environmental bodies, and the corporate and business world.

Given these goals, why is it that parts of the geographic profession display such distrust of the developments in GIS and remain skeptical about the motives, potential value, and political consequences of its adoption?

If the modernizing impulse of electronic technology is interpreted by some as liberating—as creating new opportunities for civil society to forge "communities" of correspondence, such as through the emergence of computerized e-mail networks and bulletin boards within universities and large corporations—others are more sanguine about the rationalizing effects of such modernizing technologies. The new systems of knowledge engineering raise many questions about freedom, civil society, and democratic practice, whether, for example, from a Marxian analysis of the differential impacts of technology adoption across race, class, and gender, a Weberian interpretation of technological modernization as part of a broader rationalization of social life, a Habermasian critique of instrumental rationality and the colonization of the lifeworld, or a Foucaultian account of the normalizing effects of new power/knowledge practices embedded in the discursive and nondiscursive practices of computerization.

As Hall (1993, p. 369) reminds us, "Reading a map represents a profound act of faith. Faith in the mapmaker, in technologies of measurement (and the science that underlies them), in the idea of the map—that the unique mosaic of boundaries and symbols corresponds to real space in what we like to call the real world." Yet the map and mapmaker have often been implicated in profound acts of betrayal:

> With centuries of distance and historical hindsight, we can see that error and bias, exploitation and colonialism, self-serving centrism and ecological harm can so easily be read into the subsoil of old maps and that they may as well be listed with symbols and explained in the legend. (Hall, 1993, p. 370)

> It would be foolish to ascribe that unspeakable tragedy [the Great Dying of the New World brought about by Spanish conquest] . . . to the maps that charted the New World; but it would be equally foolish to ignore the intricate weave of social and cultural nerves that connect discovery, exploration, and mapmaking. The map is the game board upon which human destinies are played out, where winning or losing determines the survival of ideas, cultures, and sometimes entire civilizations. (Hall, 1993, p. 371)

Hall (1993) is refreshingly clear-sighted about the exciting possibilities of new maps and their inherent dangers:

> Every map presages some form of exploitation. . . . Geopolitics, after all, is impossible without a cartographer, and that exercise of control over a distant domain marks a watershed in political power, confirming the notion that maps are not merely pictures of the world, but depictions of a world that can be shaped, manipulated, acted upon. (p. 383)

> Map historian J. B. Harley refers to cartography as the "science of princes," and it is a characterization that applies to modern mapmakers as well. From the expeditions financed by Spain and Portugal in the fifteenth century to experiments sponsored by the National Science Foundation last year, there exists a tradition of what might be called "mercenary geographers." In the context of contemporary science, the term strikes the ear harshly; but in the context of the history of exploration and mapping, there is compelling and overwhelming evidence that "explorers," terrestrial and intellectual, must align their professional and personal ambitions with wealthy and powerful nations, which can afford the expeditions (or, in the modern analogue/idiom, the "experiments") that chart and stake a claim to new territories. (p. 384)

Thus, Hall (1993, p. 387) asks, "Can we acquire modern map knowledge without . . . inventing and committing new, equally modern and unimagined cruelties?"

This question becomes even more pertinent with electronic systems of representation (such as GIS), where the pace and scope of adoption, and the at times unsavory uses to which the information and technology has been put (and for which it was, in part, developed), have encouraged only limited challenges to instrumental conceptions of the role of GIS in society (see, e.g., Pickles, 1991; Smith, 1992; O-Tuathail, 1993). But, as GIS has become—along with hypertext, multimedia, and other complex, multilevel computer database and imaging systems—an element in the extension of accounting systems and the servicing of new needs and responsibilities on the part of public and private agencies, it becomes crucial to ask how these technologies impact on the ways in which people interact with one another.

Despite the apparently pressing nature of these issues, discussion of the social impacts of GIS has been limited mainly to an internal analysis of technique and practice. Little external evaluation and critique has been developed. Where this has occurred, the heat and emotion surrounding issues of reallocation of funding, redirection of teaching and research programs, and competition between GIS and other areas of the discipline have tended to result in angry polemic, instead of thoughtful, strong theoretical engagement.[7] Of course, since the personal, institutional, and social stakes are high, this is not unexpected. The development and adoption of new information technologies, and the rise of new ways of doing things, do not occur without struggle. Each new technology, insofar as it is widely adopted, must be situated within existing norms, practices, and discourses, or new norms, practices, and discourses must be created. This is no less true for the electronic media of remote surveillance and multiscale mapping technologies that lie at the core of GIS.

The discussion of GIS in geography seems to have taken one of two tacks. Either GIS is interpreted from within the ranks of the practice itself, in which case interpretations reflect a concern for questions of method and technique. Or geographers have become advocates of GIS—the new "space cadets." Yet many seem unwilling to discuss the destabilizing effects on GIS of even the most pertinent recent debates about the sociospatial dialectic, power/knowledge, and the constitution of social and political subjects. Instead, much discussion takes the form of unreflective GIS advocacy and an almost evangelical need to proselytize about the geographical nature of GIS. Objectivist epistemology and a pragmatic politics combine to reject any broader theorizing of the consequences of this form of knowledge production and management.

In what ways can a social theoretic understanding transcend these polar positions, and how can we speak about this technology without presupposing the ontological and epistemological assumptions on which GIS is founded? For many of the authors in this book the question is not only about the internal possibilities and constraints of GIS, but about the reconfigurations of social, economic, political, and disciplinary life that the emergence of electronic technologies like GIS are creating. However, one of the central difficulties in developing a critical social theory of GIS is the refusal of GIS users to distinguish between empirical and technical claims about objects, practices, and institutions, and the discourses within which particular claims about objects, practices, and institutions, and particular claims to truth, are made. That is, concepts, practices, and institutional linkages remain largely unproblematized, naturalized as normal and reasonable ways of thinking and acting.

The language in use in GIS itself is instructive. In the words of GIS exponents and practitioners the new electronic technologies permit the rapid and extensive *surveying* of *new and more complete sets of data* at great *speed,* decreasing *cost,* and greater *efficiency.* The *technological* changes that make these *advances* possible also permit the *standardization* and *manipulation* of a variety of discrete data sets to *yield* new *spatially specific* sets of information that can be *codified,* and even *commodified.* This *control technology* and *knowledge engineering* require special *skills, knowledge,* and *training.* The *output* is in great *demand,* students can find good *jobs,* and government, military, and business *applications* provide challenges for the university researcher.

These claims are made, however, in the almost total absence of any wider context of theorizing the changes in technology and social relations, of epistemology and theories of science, or of the processes of the production, representation, and dissemination of information within which these processes operate. As Foucault (1980) has so clearly demonstrated, the technics of the human sciences have arisen in conjunction with specific practices in the broader society:

> In a society such as ours, but basically in any society, there are manifold relations of power which permeate, characterize and constitute the social body, and these relations of power cannot themselves be established, consolidated nor implemented without the production, accumulation, circulation and functioning of a discourse. There can be no possible exercise of power without a certain economy of discourses of truth which operates through and on the basis of this association. We are subjected to the production of truth

> through power and we cannot exercise power except through the production of truth. (p. 93)

The task of a critical genealogy of power is to clarify the detailed practices that constitute the "history of the present," and to provide accounts of the emergence of new modalities of power (Fraser, 1989, p. 17).

GIS is just one of these new complexes of discourse, practice, and institutional ensemble, among many others, effecting changes in the modalities of power. As a cultural practice, instituted historically, its forms and effects are consequently contingent, ungrounded except in terms of other, prior, contingent historically instituted practices. In this sense, power is as much about the possibilities of modernization—the ways in which identity and difference are constituted—as about the exercise of influence and the formation of new iron cages. As social relations and new subjectivities are embodied, we need to ask how such identities are sustained, how power flows through the capillaries of society in particular settings, and what role new technologies of the self and of society play in this circulation of power. Foucault would have understood well our contemporary fascination with GIS, its "technologies of surveillance," forms of knowledge engineering, and commitment to the categorizing and normalizing of nature and social life.

In *The Consequences of Modernity* Giddens (1990) raised the question of modernity and trust: As more and more people live in situations in which they interact with disembodied institutions, their local practices are mediated by globalized social relations, and their daily contacts are increasingly mediated by automated and computerized operations (the bank teller machine, automatic telephone answering machines, camera operated security systems, etc.); *facework commitments* (which "are sustained by or are expressed in social connections established in circumstances of copresence" [p. 80]) are replaced by *faceless commitments* (which depend upon "the development of faith in symbolic tokens or expert systems" [p. 80] or abstract systems). The shifting balance between facework and faceless commitments and contacts is also a recomposition of the rules and practices that constitute social behavior. Thus, for example, the complex skills required to maintain civil inattention (the form of encounter that takes place between strangers in a community) are replaced by alternative systems of encounter, such as forms of uncivil inattention like the hate stare. The possibilities for deep-seated changes in the nature of social life are very real in such abstract systems where the nature of trust and interaction change.

CONCLUSION

> *What, in a positive sense, made the new communities imaginable was a half-fortuitous, but explosive, interaction between a system of production and productive relations (capitalism), a technology of communications (print), and the fatality of human linguistic diversity. (Anderson, 1983, p. 43)*

In the debate about the nature, uses, and impacts of GIS in the hypermodern world of generalized information and communication, geographers have adopted a relatively limited range of critical positions. For many, GIS offers exciting professional opportunities and personal challenges. For others, GIS represents a reassertion of instrumental reason in a discipline that has fought hard to rid itself of notions of space as the dead and the inert, and, as Soja (1989) has argued, to reassert a critical understanding of the sociospatial dialectic. For yet others, the debate about GIS is a nonissue (Clark, 1992). As I hope I have begun to show in this first chapter, the emergence of GIS as both a disciplinary practice and a socially embedded technology represents an important change in the way in which the geographical is being conceptualized, represented, and materialized in the built environment. As both a system for information processing and for the creation and manipulation of spatial images, and as a technology which is diffusing rapidly through the apparatuses of the state and the organs of business, GIS requires a critical theory reflecting sustained interrogation of the ways in which the use of technology and its products reconfigure broader patterns of cultural, economic, or political relations, and how, in so doing, they contribute to the emergence of new geographies.

Along with the important critical task of assessing the impacts of GIS as tool, technology, and social relation, I have also tried to show how we need to think more seriously about the transformative possibilities that GIS offers. In regard to parallel developments in cyberspace, Heim (1992, p. 59) has argued that "cyberspace is more than a breakthrough in electronic media or in computer interface design. With its virtual environments and simulated worlds, cyberspace is a . . . tool for examining our very sense of reality." Whether and how this tool for manipulating and understanding both our world and our sense of our world is used, depends at least in part on the conceptual tools, critical frameworks, and linguistic codes we choose to mobilize by way of response.

The chapters in this book seek to contribute to this task of theorizing GIS. Each author seeks to challenge the auratic character of the technology and its uses, suspend acceptance of legitimating claims made about GIS based solely on a naturalized notion of progress, describe the broader impacts GIS and related information and imaging technologies are having, and develop a range of vocabularies and critical perspectives with which to approach these issues.[8]

NOTES

1. For a critique of the assumptions behind this view of mapping, see Wood (1992).

2. Of course, all these matters are contingent on the types of regulatory framework that emerge to govern development, property rights, access, and so on.

3. For a critique of paradigmatic thinking and an argument for post-paradigmatic science, see Pickles and Watts (1992).

4. This discussion of Martin is based on Pickles (1992a).

5. Several conceptions of representation and reality underpin this understanding of ''socioeconomic applications.'' GIS is concerned with the representation of spatial data. Such representational practices are made distinct by the ''ways in which data are organized in GIS to provide a flexible model of the real world'' (Martin, 1991, p. 8). These new computer-generated representations—''virtual maps''—are distinct from ''real maps'' in that they offer a greater degree of flexibility (Martin, 1991, p. 13). Martin does stress the filtering effect of this represenational act, suggesting that all remotely sensed images are but poor representations of the real world (p. 22), and that classification systems may bear little relationship ''to the 'real world' classes of land cover which we are hoping to discover'' (p. 23). The ''real'' here is that which is naturally given in unmediated form: a land surface to be captured as a raw image to be classified (p. 21). The task of the GIS user is to represent and manipulate a model of geographic reality as accurately as possible (p. 27).

6. See Dreyfus (1992) for critical reflections on this issue, Fontaine (1992) for an uncritical example, and Dobson (1983, 1993), Pickles (1993), and Sheppard (1993) for reflections on this issue in geography.

7. See the polemics in Oppenshaw (1991, 1992) and Taylor and Overton (1991).

8. By ''critical perspective,'' of course, I do not mean merely direct criticism of GIS, but a critical analysis of the effects brought into play in the dis-

ciplines, particular institutions, and society generally when GIS is deployed as a pedagogic or research tool, as a system of accounting, and as a system of control.

REFERENCES

Abler, R. A. (1993). Everything in its place: GPS, GIS, and geography in the 1990s. *Professional Geographer, 45*(2), 131–139.

Abler, R. F., Adams, J. S., & Gould, P. R. (1971). *Spatial organization: The geographer's view of the world.* Englewood Cliffs, NJ: Prentice-Hall.

Anderson, B. (1983). *Imagined communities: Reflections on the origins and spread of nationalism.* London: Verso.

Barry, D. (1991, October). Fleet management makes advances with digital mapping technology. *GIS World, 4*(10), 74–77.

Barry, J. (1992). Mappings—A chronology of remote sensing. In J. Crary & S. Kwinter (Eds.), *Incorporations* (pp. 570–571). Cambridge, MA: MIT Press.

Benedikt, M. (1991). *Cyberspace: First steps.* Cambridge, MA: MIT Press.

Benjamin, W. (1968). The work of art in the age of mechanical reproduction. In H. Arendt (Ed.), *Illumination: Essays and reflections,* (pp. 217–257). New York: Schocken Books.

Clark, G. L. (1992). Commentary: GIS—what crisis? *Environment and Planning A, 23,* 321–322.

Clark, R. E. (1991, October). GIS solves Denver car dealer's dilemma. *GIS World, 4*(10), 67–73.

Clarke, K. C. (1986). Advances in geographic information systems. *Computers, Environment and Urban Systems, 10,* 175–184.

Clarke, S. (1988). Overaccumulation, class struggle and the regulation approach. *Capital and Class, 36,* 59–92.

Cooke, D. F., & Plumer, C. (1991, October). *GIS World, 4*(10), 82–85.

Coppock, J. T., & Rhind, D. W. (1991). The history of GIS. In D. J. Maguire, M. F. Goodchild, & D. W. Rhind (Eds.), *Geographical information systems* (Vol. 1, pp. 21–43). London: Longman Scientific and Technical/New York: Wiley.

Crary, J., & Kwinter, S. (Eds.). (1992). *Incorporations.* Cambridge, MA: MIT Press.

Dangermond, J. (1991, October). Business adapting GIS to a host of applications. *GIS World, 4*(10), 51–56.

Derrida, J. (1976). *Of grammatology* (G. C. Spivak, Trans.) Baltimore: Johns Hopkins University Press.

Dobson, J. (1983). Automated geography. *Professional Geographer, 35*(2), 135–143.

Dobson, J. (1993). The geographic revolution: A retrospective on the age of automated geography. *Professional Geographer, 45*(4), 431–439.

Dreyfus, H. L. (1992). *What computers still can't do: A critique of artificial reason.* Cambridge, MA: MIT Press.

Fontaine, G. (1992). The experience of a sense of presence in intercultural and international encounters. *Presence, 1*(4), 482–490.

Foucault, M. (1980). Two lectures. In C. Gordon (Ed. and Trans.), *Power/knowledge: Selected interviews and other writings, 1972–1977* (pp. 78–108). New York: Pantheon Books.

Francica, J. R. (1991, October). GIS in business. *GIS World, 4*(10), 49.

Fraser, N. (1989). *Unruly practices: Power, discourse, and gender in contemporary social theory.* Minneapolis: University of Minnesota Press.

Gatrell, A. (1983). *Distance and space: A geographical perspective.* Oxford: Clarendon Press.

Gelernter, D. (1992). *Mirror worlds or the day software puts the university in a shoebox . . . How it will happen and what it will mean.* New York: Oxford University Press.

Giddens, A. (1990). *The consequences of modernity.* Stanford, CA: Stanford University Press.

Gusso, J. S., & Lasala, V. (1991, October). Desktop mapping delivers GIS efficiency to telecommunication. *GIS World, 4*(10), 86–88.

Hall, S. S. (1993). *Mapping the next millennium: How computer-driven cartography is revolutionizing the face of science.* New York: Vintage Books.

Harley, B. (1988a). Silences and secrecy: The hidden agenda of cartography in early modern Europe. *Imago Mundi, 40,* 57–76.

Harley, B. (1988b). Maps, knowledge and power. In D. Cosgrove & P. Daniels (Eds.), *The iconography of landscape* (pp. 227–312). Cambridge: Cambridge University Press.

Harley, B. (1989). Deconstructing the map. *Cartographica, 26*(2), 1–20.

Harley, B. (1990). Cartography, ethics and social theory. *Cartographica, 27*(2), 1–23.

Harvey, D. (1989). *The condition of postmodernity: An inquiry into the origins of cultural change.* Oxford: Basil Blackwell.

Heidegger, M. (1977). The question concerning technology. In W. Lovitt (Ed. and Trans.), *The question concerning technology and other essays* (pp. 3–35). New York: Harper Colophon.

Heim, M. (1992). The erotic ontology of cyberspace. In M. Benedikt (Ed.), *Cyberspace: First steps* (pp. 59–80). Cambridge, MA: MIT Press.

Heivly, C. (1991, October). Route planning tool boosts saturation delivery efficiency. *GIS World, 4*(10), 78–79.

Huxhold, W. E. (1991, March). The GIS profession: Title, pay, qualifications. *Geo Info Systems, 12,* 22.

Keating, R. (1992, May). Building accuracy into GIS. *GIS World, 5*(5), 32–34.

Kirchner, R., & Thomas, R. K. (1991, October). Dunkin' Donuts plugs hole in location strategy. *GIS World, 4*(10), 89.

Lake, R. W. (1993). Planning and applied geography: Positivism, ethics, and geographic information systems. *Progress in Human Geography, 17*(3), 404–413.

Landow, G. P. (1992). *Hypertext: The convergence of contemporary critical theory and technology.* Baltimore: Johns Hopkins University Press.

Luke, T. W. (1993, April 7–9). *Beyond Leviathon, beneath Lilliput: Geo-*

politics and glocalization. Paper presented at the annual meeting of the Association of American Geographers, Atlanta, GA.
Maffini, G. (1991, October). GIS at threshold of business applications. *GIS World, 4*(10), 50–52.
Maguire, D. J., Goodchild, M. F., & Rhind, D. W. (1991). *Geographical information systems* (Vol. 1). London: Longman Scientific and Technical/New York: Wiley.
Mahon, R. (1987). "From Fordism to ?" New technology, labour markets and unions. *Economic and Industrial Democracy, 8,* 5–60.
Martin, D. (1991). *Geographic information systems and their socioeconomic applications.* Andover, Hants, England: Routledge, Capman and Hall.
Moloney, T., & Dellavedova, B. (1991, October). Retail applications reap the benefits of GIS. *GIS World, 4*(10), 62–66.
Moncla, A., & McConnell, L. (1991, October). TIGER helps retailers address marketing issues. *GIS World, 4*(10), 80–82.
Murdock, G., & Golding, P. (1989). Information poverty and political inequality: Citizenship in the age of privatized communications. *Journal of Communication, 39,* 180–196.
Nelson, T. H. (1992). Virtual world without end: The story of Xanadu. In L. Jacobson (Ed.), *Cyberarts: Exploring art and technology* (pp. 157–169). San Francisco: Miller Freeman.
Newell, R. G., & Theriault, D. G. (1990). Is GIS just a combination of CAD and DBMS? *Mapping Awareness, 4*(3), 42–45.
Norris, C. (1987). *Derrida.* Cambridge, MA: Harvard University Press.
Olsson, G. (1972). Some notes on geography and social engineering. *Antipode, 4*(1), 1–21.
Olsson, G. (1974). Servitude and inequality in spatial planning: Ideology and methodology in conflict. *Antipode, 6*(1), 16–21.
Olsson, G. (1992). *Lines of power. Limits of language.* Minneapolis: University of Minnesota Press.
Openshaw, S. (1991). Commentary: A view on the GIS crisis in geography, or, using GIS to put Humpty-Dumpty back together again. *Environment and Planning A, 23,* 621–628.
Openshaw, S. (1992). Commentary: Further thoughts on geography and GIS: A reply. *Environment and Planning A, 24,* 463–466.
O'Tuathail, G. (1993). The new East–West conflict? Japan and the Bush administration's "New World Order." *Area, 25*(2), 127–135.
Penley, C., & Ross, A. (Eds.). (1991). *Technoculture.* Minneapolis: University of Minnesota Press.
Pickles, J. (1985). *Phenomenology, science, and geography: Space and the human sciences.* Cambridge: Cambridge University Press.
Pickles, J. (1986). *Geography and humanism.* CATMOG, *44.* Norwich, England: Geo Books.
Pickles, J. (1991). Geography, GIS, and the surveillant society. *Papers and Proceedings of Applied Geography Conferences, 14,* 80–91.
Pickles, J. (1992a). Review of "Geographic information systems and their socio-

economic applications." *Environment and Planning D: Society and Space, 10,* 597–606.

Pickles, J. (1992b). Texts, hermeneutics, and propaganda maps. In T. J. Barnes & J. S. Duncan (Eds.), *Writing worlds: Discourse, text, and metaphor in the representation of landscape* (pp. 193–230). New York: Routledge.

Pickles, J. (1993). Discourse on method and the history of discipline: Reflections on Jerome Dobson's 1993 "Automated geography." *Professional Geographer, 45*(4), 451–455.

Pickles, J., & Watts, M. (1992). Paradigms of inquiry? In R. F. Abler, M. G. Marcus, and J. M. Olson (Eds.), *Geography's inner worlds: Pervasive themes in contemporary American geography* (pp. 301–326). New Brunswick, NJ: Rutgers University Press.

Poster, M. (1990). *The mode of information: Poststructuralism and social context.* Cambridge, England: Polity Press.

Pred, A., & Watts, M. J. (1992). *Reworking modernity: Capitalisms and symbolic discontent.* New Brunswick, NJ: Rutgers University Press.

Rheingold, H. (1992). *Virtual reality.* London: Mandarin.

Ronnel, A. (1989). *The telephone book: Technology, schizophrenia, electric speech.* Lincoln: University of Nebraska Press.

Rosenthal, P. (1992). Remixing memory and desire: The meanings and mythologies of virtual reality. *Socialist Review, 22*(3), 107–117.

Runnels, D. (1991, October). Geographic underwriting system streamlines insurance industry. *GIS World, 4*(10), 60–62.

Sheppard, E. (1993). Automated gography: What kind of geography for what kind of society? *Professional Geographer, 45*(4), 457–460.

Smith, N. (1992). Real wars, theory wars. *Progress in Human Geography, 16*(2), 257–271.

Soja, E. (1989). *Postmodern geographies: The reassertion of space in critical social theory.* London: Verso.

Taylor, P. (1990, July). Editorial comment: GKS. *Political Geography Quarterly, 9*(3), 211–212.

Taylor, P., & Overton, M. (1991). Commentary: Further thoughts on geography and GIS. *Environment and Planning A, 23,* 1087–1094.

Wood, D. (1992). *The power of maps.* New York: Guilford Press.

Wooley, B. (1992). *Virtual worlds: A journey in hype and hyperreality.* Cambridge, MA: Blackwell.

Zuboff, S. (1984). *In the age of the smart machine.* New York: Basic Books.

CHAPTER 2

Geographic Information Systems and Geographic Research

Michael F. Goodchild

Many disciplines have contributed to the development of geographical information systems (GIS), and in turn GIS has been used in many disciplines as a research tool, but there is no doubt that GIS and geography have a special relationship. This chapter explores some of the dimensions of that relationship, with particular emphasis on geographic research.

Good debate is entertaining. Because I am the author in this book most clearly identified with GIS, many readers, I suspect, are hoping to be entertained by my clever defense of GIS, and perhaps even by my rousing counterattack to what they may interpret in recent literature as critique. Phrases like "GIS über alles" (Smith, 1992) have appeared in the pages of geography's more respected journals, and a colleague has written that its "basic goals . . . are precisely to foster the technics and ideology of normalization" (Pickles, 1991, p. 83). From the other side, Openshaw's (1991) editorial in *Environment and Planning A* was certainly a spirited defense of GIS, as was his later response (Openshaw, 1992) to Taylor and Overton (1991).

But Smith (1992) is perfectly correct in pointing out that the GIS literature places far more emphasis on civilian applications and tends to ignore—or to be ignorant of—the military ones, and Pickles is correct too when he points out that GIS can be used in support of civilian surveillance. As academics, it is our responsibility to reflect on all aspects of GIS, from its basic design and functionality to the more profound aspects of its meaning to society. While it is hard to see power in the possession of a soil map, or politics in the measurement of atmospheric temperature, there are real ethical issues arising from many applications of GIS: a technology that can be used to promote democracy can also be used to deny it. The gerrymandered 1992 electoral map of North Carolina (see Figure 2.1) was designed by a GIS to empower minorities, but previous generations would have seen creation of such an engineered district as an extreme abuse of the electoral process. Another GIS product (see Figure 2.2), prepared by Lauretta Burke, compares the locations of industries emitting toxic chemicals into the atmosphere of Los Angeles with the locations of census tracts occupied primarily by minority populations. It makes a powerful statement of spatial association, and played a minor role in the 1992 election campaign.

The role played by GIS in society is clearly an important dimension of the relationship between GIS and geographic research, and possibly the most important in many contexts, but it is only one dimension. Every student of GIS should be aware of the technology's possible uses,

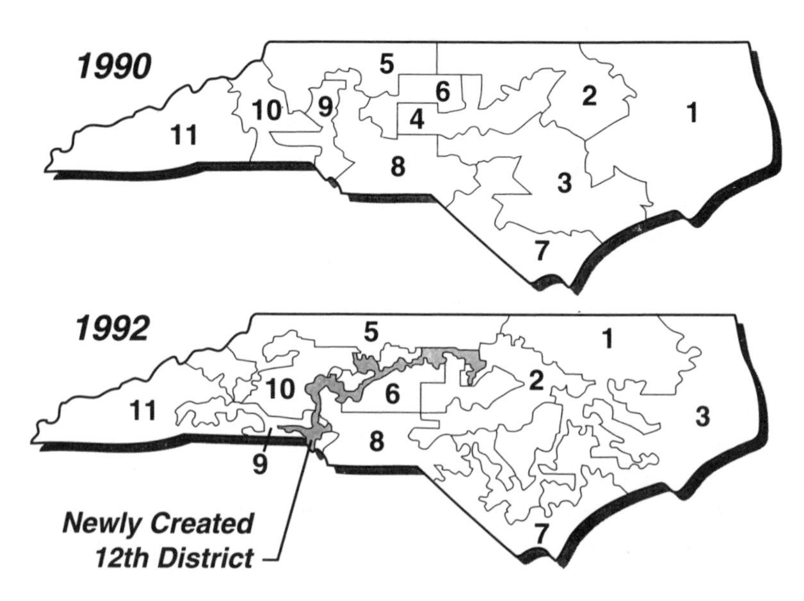

FIGURE 2.1. Congressional districts formed by the 1992 reapportionment of North Carolina, with 1990 districts for comparison.

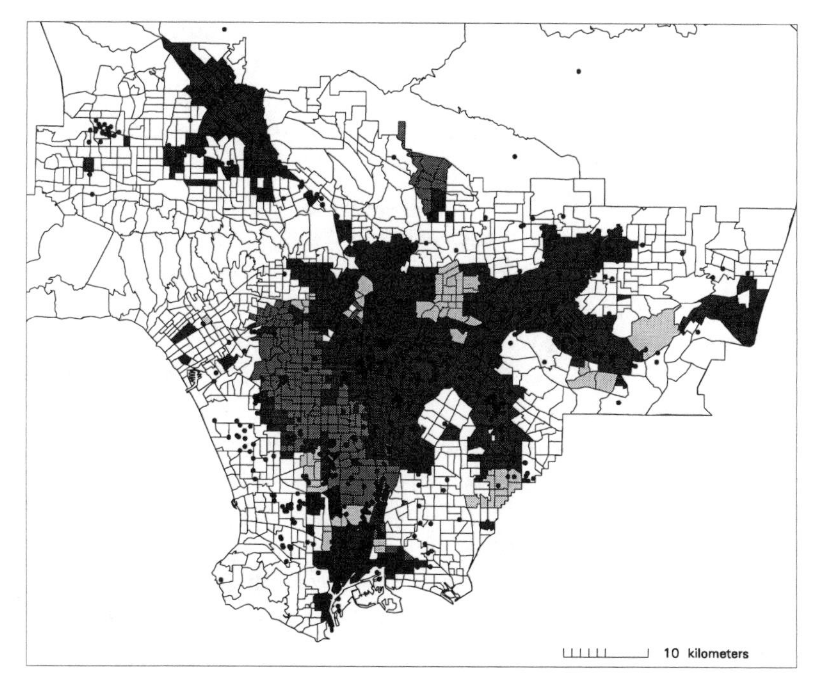

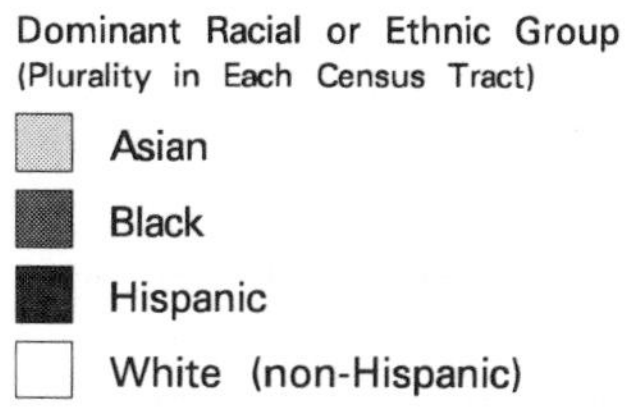

FIGURE 2.2. Locations of industries emitting toxic chemicals (data source: 1989 U.S. Environmental Protection Agency, Toxic Release Inventory database), and dominant ethnic group by census tract (source: U.S. Bureau of the Census, 1990 census), for Los Angeles County (Burke, 1993).

for both good and evil, and of the difficulties we all frequently face in making such clear distinctions, even with the benefit of hindsight. But writers often take extreme positions in debate, perhaps to be more exciting, or because they feel driven to present their position forcefully in order to counter what they see as incorrect tradition. It seems absurd to me to suggest that all users or developers of GIS should see their relationship to the Persian Gulf War (Smith, 1992) as somehow similar to Robert Oppenheimer's moral dilemma over the atomic bomb, or to argue that GIS is a "surveillance technology" (Pickles, 1991), or conversely to argue that the white knight of GIS will put geography's pieces together again (Openshaw, 1991). While this kind of debate may be fun for a while, there are nevertheless serious and interesting issues to be resolved, and scientific questions to be pursued in GIS, and it would be a shame if they were lost in a cacophony of position taking. I hope the reader will be convinced, by the end of this chapter, that both GIS and geography are relatively small specializations, and each needs the other badly if both are to survive.

In the following sections I will attempt to review the relationship between GIS and geographic research, in all of its dimensions. The chapter begins with an overview of GIS technology, and then moves to a discussion of research *on* GIS, and the nature of the GIS research community. The following section discusses research *with* GIS, and the role that can be played by geographers in that multidisciplinary activity. The final section looks briefly at the ethics of GIS and its significance to society, and ends with a call for extensive research on the topic by geographers.

Before I begin, it seems appropriate to provide a context for this chapter. If GIS must be examined in its societal context, then so too must writing about GIS, so perhaps a little self-deconstruction is in order. My undergraduate degree was in physics and my Ph.D. in physical geography, and I still find myself able to write words like *scientific, objective,* and *truth* without placing them in quotation marks, and to associate the name "Foucault" with a pendulum. While I would agree that social science often says as much about the observer as about the observed, I have seen little in the way of alternatives to positivism among the more fundamental physical sciences. Though much of GIS literature can be rightly criticized for ignoring military applications, it seems to me that much writing in geography about the decline of positivism similarly ignores the physical side of the discipline. Of course, part of the attraction of GIS is its ability to surmount the human/physical divide. So unlike Heywood (1990), Taylor (1990), and others who see in GIS a resurgence of positivism, I see the GIS literature as containing elements of the entire spectrum, from the positivist end to the other end. Cartographers writing on GIS sometimes object strongly to words like *truth,* whereas computer scientists seem equally impatient with *debate.*

But however strong the case against positivism in the social sciences may be, I also believe that quantification, analysis, numerical models, and related concepts provide us with valuable points of reference. Objectivity is always embedded in the subjectivity of human experience, and the success of disciplines like physics lies in their ability to extend objectivity over such a large, connected set of phenomena. In the social sciences, the extent of objectivity is much more limited, and the subjective context is encountered more immediately. But however small, islands of objectivity—the spatial interaction model, central place theory, microeconomics—provide us with the well-defined points of reference that make debate and intellectual progress possible.

THE NATURE OF GIS

Though many definitions of GIS exist, most identify a database in which every object has a precise geographical location, together with software to perform functions of input, management, analysis, and output. Besides geographical locations, the database will also contain numerous attributes that serve to distinguish one object from another, and information on the relationships between objects. There are several excellent introductions to GIS, including Burrough (1986), Star and Estes (1990), and Thompson and Laurini (1992). Maguire, Goodchild, and Rhind (1991) provide a comprehensive overview of GIS.

This section provides three different perspectives on the current nature of GIS: as a technology, as a research field, and as a community. All three are legitimate interpretations of what is meant by the acronym, and yet together they contain an enormous variety of activities, personalities, and capabilities. In some ways the success of GIS is related to the power of labels in society—as an acronym, it is initially free of associations, and can act as a rallying point in ways that traditional terms overburdened with meaning, like *geography,* cannot.

GIS: The Technology

Although computer hardware and specialized peripheral equipment are essential for GIS, the key component of this particular technology is its software. Over the past decade, interest in GIS has provided a significant incentive to developers, and the range of products that call themselves GIS has grown wider and wider. A 1991 directory (GIS

World, Inc., 1991) listed 371 available software products, representing an enormous diversity of capabilities and approaches. The borders of GIS are very fuzzy, particularly with remote sensing, computer assisted design (CAD), and computer cartography, all of which are recognized areas of software that to some degree meet the definition of GIS given above. Various tests have been suggested or applied in an effort to refine the definition, including the following:

- The ability to store and analyze spatial relationships between objects, such as *crosses, intersects, is adjacent to,* or *is connected to,* or to compute them as required (often called *topology* in the loose terminology of the GIS community)
- The ability to store and analyze an unlimited number of attributes of each object
- An emphasis on analysis, rather than simple data management and retrieval
- The ability to integrate data from different sources, perhaps at different scales and using more than one mode of representation

The most effective principle for organizing the range of GIS software is based on each product's underlying *data model.* In computer science, a data model is the set of rules used to create a representation of information, in the form of discrete entities and the relationships between them. Thus geographical data modeling (Goodchild, 1992a; Peuquet, 1984) is the set of rules used to create a representation of geography in the discrete, digital world of a computer database. The human mind uses a myriad of poorly understood methods for structuring geographical knowledge; it is GIS's supreme conceit that one can structure a useful representation of geographical knowledge in the absurdly primitive domain of the digital computer, just as it is cartography's conceit that one can accomplish the same objective with pen and paper. Yet clearly there are areas of human activity—finding underground pipes, tracing the ownership of land, navigating through unfamiliar cities, managing forests—where it can be done with satisfaction. As Taylor (1990) has pointed out, it is much easier to do so when the information being modeled consists of geographical facts (bridges, streets, buildings) than when it consists of geographical interpretations of complex phenomena, like soil, terrain, or urban landscapes, or of geographical knowledge and understanding. Hence the danger arises that a geography that accepts GIS too readily will become a discipline dominated by facts rather than by understanding.

Two classes of data models have dominated GIS over the past decade, although one arrived much later than the other. The roots of GIS lie in the 1960s, and it is generally acknowledged that the first GIS was the Canada Geographic Information System (Tomlinson, 1988), developed by and for the Government of Canada to support the mapping and assessment of Canada's land base. Land was inventoried in the form of a number of distinct variables—capability for agriculture, capability for recreation, land use, and so on—and assembled into maps, initially at a scale of 1:50,000. Since each variable was determined uniquely at every point, the maps could be conceptualized as a series of *layers,* or *fields* in mathematical terms, and the database as a layer cake. Thus the distinct feature of the field class of data models is a database that contains a finite number of variables, each mapped over the area covered by the database, and each having a unique value at every point in the area.

The second significant root of GIS lay in the U.S. Bureau of the Census and its management of the 1970 census. Here again, fields provided an acceptable conceptual framework for data modeling, since every point in the United States is in exactly one state, one county, one census tract, or the like. When these two threads came together in the work of the Laboratory for Computer Graphics at Harvard in the 1970s (Chrisman, 1988) the layer-cake view of the world was set to dominate GIS. The ARC/INFO GIS, developed and marketed by Environmental Systems Research Institute (ESRI) and one of today's most successful GIS, is the direct intellectual descendant of that work at Harvard, as are several other current products.

There are many ways of representing a field as a collection of discrete objects, and GIS currently makes extensive use of six of them. One can sample the field at randomly located points (e.g., weather stations), or at a grid of regularly spaced points. One can divide the space into rectangular cells, and record the average, total, or dominant value in each cell (e.g., remote sensing). One can divide the space into areas that are more or less homogeneous and record the average, total, or dominant value in each area (e.g., census data, or soil maps). One can record the locations of lines where the field has certain fixed values (e.g., contour or isopleth maps). And finally, one can divide the space into irregularly shaped triangles, and assume that the field varies linearly in each (the TIN, or triangulated irregular network model, used commonly to model surfaces, e.g., topography).

All of these alternative field models are approximations, made in the interests of capturing a reasonably accurate representation of a complex phenomenon. The quality of the approximation is clearly of concern, and there are difficult choices to be made in ensuring that the representation is as accurate and useful as possible. Such choices are best made by people who understand the phenomenon and the processes that caused its particular geographic distribution. Ideally, a digital representation of a complex geographic field should capture elements of our knowledge and understanding of the phenomenon, through choices about what to measure, where to measure it, and how to represent the results in digital form. Just as a soil map captures the training, knowledge, and understanding of the soil scientist that made it, so too should a GIS representation, but without some of the constraints imposed by cartographic technology (Goodchild, 1988). All too often, however, the same choices are constrained by software limitations, or by lack of understanding on the part of the GIS user.

No current GIS gives its users full access to all six field data models. So-called *raster* GISs support only grids of regularly spaced points and rectangular arrays of cells, and do not distinguish between them. Each layer in the database of such systems must have identical size, spacing, and orientation, so that cells on one layer match perfectly with cells on all other layers. Well-known raster GISs include IDRISI, an excellent PC-based system developed at Clark University; GRASS (Geographic Resources Analysis Support System), a Unix workstation-based system developed by the U.S. Army Corps of Engineers and widely adopted in federal agencies for environmental and resource management; and MAP-II (Map Analysis Package), developed for the Macintosh II by Micha Pazner, now of the University of Western Ontario, and distributed by Wiley.

So-called *vector* GISs, on the other hand, support representation of fields through irregularly spaced points, irregular areas, irregular triangles, or contour lines. Each is regarded as a collection of objects—points, lines, and areas, respectively—with associated attributes. The geometric forms of areas and lines are normally represented as points (pairs of coordinates) connected by straight lines. Thus areas are often called *polygons* in GIS terminology, and lines are often called *polylines* by extension.

The second class of data models takes a very different approach. Since 1980, the most rapidly developing and largest area of GIS application has been in local government and utility companies. Far from seeing the world as a layer cake of fields, these applications are dominated by a view of the world as an empty space populated by various kinds of discrete objects. A telephone company must manage a vast and complex network of facilities, including poles, connection boxes, cables, and so on, each of which can be regarded as a discrete and well-defined object. The space in between the objects is empty, and there is no value in using space in the computer by creating representations of it, such as empty cells. For this reason, and also because such systems typically need high spatial resolution, they are exclusively handled by vector GIS. The term *layer* may be used, but it has a much looser meaning than for field data models, and is merely employed to group together collections of objects for management purposes. In an object model, any place may be empty or occupied by one or more objects, in one or more layers, whereas in a field model every place has exactly one value on every layer. The terms *coverage* and *theme* are broadly synonymous with *layer*.

The objects used to represent a field—points, lines, or areas—must satisfy certain constraints: Contours cannot cross, and areas representing a field cannot overlap. Thus although field models and object models are both represented internally as collections of points, lines, and areas, their behaviors and meanings are very different. In a *layer-based* vector GIS, such as ARC/INFO, polygons in one coverage must follow the rules of a field, that is, they must exhaust the space and not overlap. In an *object-based* vector GIS, such as System/9 (Computervision), on the other hand, objects follow the rules of an object model: The space between them is empty and they may overlap. Recently, systems like GDS (Geographic Data System; marketed in the United States by EDS Inc.) have begun to apply the field/object distinction locally and selectively within layers of the database. Within a local government database, for example, land parcels exhaust the limits of a city block and do not overlap, but are surrounded by empty space, and may be crossed by other objects of different classes, such as creeks.

These options, which stem ultimately from different ways of viewing the occupation of geographic space, create a wide and confusing range of options for the GIS user. The absence of strong organizing principles or a rigorous terminology has meant that it is easier to be trained in the operation of one system than to be educated in the concepts of the field of GIS as a whole. It has meant that a database constructed using ARC/INFO may be of very little value to a user of System/9, or of Intergraph's TIGRIS, because of fundamental data model differences. In time these problems should resolve themselves, as the field develops a better conceptual framework and more consistent terminology, but that process will be slow.

Field models are difficult to update because of the integrity that must extend over each layer of the database. It is difficult, for example, to update or modify a contour model because of the need to preserve spatial relationships between adjacent contours. Layer-based GISs are poorly adapted for selective update or editing of layers, and consequently do not support time-dependent data well. In a remote sensing system, each old image is replaced by a complete new image, leading to redundancy wherever the image has not changed. Object-based GISs are inherently better adapted to temporal dependence, but in general GIS today remains a technology for static data, a major impediment to its use in modeling social and economic systems.

Another, and perhaps more serious, impediment to the use of GIS in social science lies in the current emphasis in its data models on the absolute positions of objects, and the inability to represent information about interaction. Couclelis (1991) has noted that this short coming has affected GIS's applications in planning, and has led to an inevitable emphasis on the physical rather than the social or the economic aspects of human activity. GIS includes much functionality for computing, storing, and analyzing the spatial relationships between objects, but has not yet addressed the need to qualify those relationships with their own attributes, such as flow, distance, or volume of trade, and to provide functions to support the display and analysis of this information, although elements have appeared in recent versions of ARC/INFO.

The final area of debate in current data models concerns the existence of hierarchical concepts in many geographical data. A map shows information in terms of one uniform scale, but a GIS database may include information drawn from many different maps, and may even present different representations of the same information. Change of scale can reveal more objects, and more detail in existing objects, and can even cause a change in the nature of an object, as when a single-line river becomes a double-line river. From a software point of view, there is often a tension in database design between the need to accommodate the hierarchical relationships between objects and the spatial relationships that exist at a single scale. Products such as ARC/INFO have resolved this tension in favor of the spatial relationships rather than the hierarchical ones, and in general current GIS technology is not good at allowing the user to represent the cross-scale structures that exist in geographical data. Only a few vector systems, such as System/9 and GDS, have implemented hierarchical concepts in their data models. On the raster side, the quadtree structure (Samet, 1990) does allow a limited form of hierarchical linkage.

If one were looking for a quick, general summary of the successes and failures of current GIS, it might look something like this:

- Two-dimensional, with some excursions into three
- Static, with some limited support for time dependence, particularly in remotely sensed imagery
- Good at capturing the physical positions of objects, their attributes, and their spatial relationships, but with very limited capabilities for representing other forms of interaction between objects
- A diverse and confusing set of data models, or general rules of spatial representation
- Still dominated by the map metaphor, or the view of a spatial database as a collection of digital maps, particularly in the first three characteristics listed above

It is perhaps remarkable given these limitations that GIS has attracted such interest, and has been adopted by so many agencies, governments, companies, and scientific researchers, who are able to find beneficial uses for the current technology despite its comparative crudity. As an industry, GIS is currently valued at some $1 billion annually, although estimates vary widely, and it has been estimated by ESRI that the U.S. annual expenditure on input of spatial data is now $4.5 billion. Clearly there is room here for much fascinating and exciting research, particularly by geographers interested in the ways people conceptualize, construct, analyze, and reason about geographical spaces.

GIS: The Research Agenda

I hope I have managed to convey in the previous section some of the difficulties that arise when one tries to make sense out of the current range of GIS products, and to suggest that challenging and fundamental research issues abound. Many of them are old issues; indeed, part of the fascination of GIS lies in the way it has remotivated interest in issues of spatial representation and cartography that have existed for centuries. Cartographers have long struggled with the difficulties of portraying interaction and change in map form; with GIS they have an opportunity to take advantage of a wealth of new technical capabilities, including animation. Monmonier (1992) has begun to explore the cartographic possibilities of what he calls "map scripts," using sequences of map and other information to convey types of messages to the user that have not been possible with traditional maps. We can now combine maps with sound and images, change scales at will, create maps from seamless databases, and generate orthographic views of three-dimensional surfaces. All of these possibilities, and more, have helped to revitalize cartography and to give an old and honored discipline new meaning.

GIS raises important issues for many disciplines, and has done much to remove the traditional isolation between photogrammetry, remote sensing, geodesy, cartography, surveying, and geography (one could add to this list computer science, operations research, spatial statistics, cognitive science, behavioral psychology, and any other discipline with interests in the generic issues of spatial data). In an earlier paper (Goodchild, 1992b) I argued that these are the disciplines of geographic information *science,* and that it made more sense for the research community to decode the GIS acronym in this way, focusing on the fundamental issues of spatial data, rather than on the limited solutions offered by today's geographic information *system* products.

There have been several published attempts to identify the set of fundamental research issues raised by GIS, and to lay them out as research agendas. These are agendas for research *on* GIS, not research *with* GIS, although clearly the first advances the second goal. There have been debates over whether such an agenda is possible, in the sense that it assumes the existence of a set of generic issues, while in reality all issues of spatial data may be specific. For example, it may be true that there are no general principles for reasoning with spatial data, and that reasoning therefore always depends on context. It may be true that uncertainty in spatial data is similarly specific to context. But GIS itself rests on the assumption that different kinds of spatial data have common structures, and are processed in similar ways, and that there is consequently value in creating common spatial data handling and processing systems. And ultimately this is one argument for the existence of geography itself as a discipline.

Maguire (1990) describes a GIS research agenda for the 1990s aimed both at advancing the technology itself and promoting our understanding of its impact on the organizations that make use of it. Better methods for assessing GIS costs and benefits are needed, as is a better understanding of its role in organizations and the factors that influence its adoption. That same theme is stressed by Craig (1989), writing on behalf of URISA (Urban and Regional Information Systems Association), whose membership includes many professional GIS users. Perhaps the most extensive discussion of GIS research topics is the agenda developed in 1987–1988 by a consortium of the University of California, Santa Barbara, the State University of New York at Buffalo, and the University of Maine, as part of a proposal to the U.S. National Science Foundation (NSF) for the National Center for Geographic Information and Analysis (NCGIA), which subsequently published the agenda in 1989. It argued that fundamental research was needed in certain specific areas to remove impediments to further development, better use, and more widespread adoption of GIS, particularly for scientific work.

Since the proposal was accepted in 1988, NCGIA research has focused on all of the five major areas originally identified by NSF:

1. Spatial analysis and spatial statistics
2. Spatial relationships and database structures
3. Artificial intelligence and expert systems
4. Visualization
5. Social, economic, and institutional impacts of GIS

Within these broad areas, research has been organized as a series of initiatives, each focused on a topic of basic scientific interest, and each lasting roughly two years:

1. Accuracy of Spatial Databases (December 1988–November 1990)
2. Languages of Spatial Relations (January 1989–July 1990)
3. Multiple Representations (February 1989–August 1990)
4. Use and Value of Geographic Information (May 1989–May 1992)
5. Design and Implementation of Very Large Spatial Databases (July 1989–June 1993)
6. Spatial Decision Support Systems (March 1990–April 1993)
7. Visualizing the Quality of Spatial Information (started June 1991)
8. Formalizing Cartographic Knowledge (started October 1993)

9. Institutions Sharing Spatial Information (started February 1992)
10. Temporal and Spatial Reasoning in GIS (started May 1993)
11. Remote Sensing and GIS (December 1990–April 1993)
12. User Interface Design (June 1991–March 1994)
13. Spatial Analysis and GIS (started April 1992)
14. Multiple Roles for GIS in Global Change Research (to start in 1994)
15. Legal Issues (to start in 1994)

Several countries have developed GIS research strategies, and there are organizations analogous to NCGIA in the Netherlands, France, Australia, and the United Kingdom.

I do not wish to imply by this focus on NCGIA that it somehow monopolizes GIS research. Rather, the center's research agenda provides a useful way of giving an overview and selection of significant research topics. A much more complete overview of international GIS research is provided by Maguire et al. (1991), or by the pages of the major journals of the field.

The issues raised by the development and use of GIS attract researchers from a number of disciplines. Although geographers continue to play a prominent role in research on GIS, borders between disciplines are comparatively unimportant, and progress on these issues will clearly require many different perspectives. In summary, GIS as a field of research is very different from the limited view offered by GIS as a technology. It is not isolated or well defined, since progress on many of these issues benefits not only GIS, but a host of related fields such as image processing, remote sensing, map production, and cognitive science. Instead, GIS as a technology is providing an essential motivation for a wide range of interesting and fundamental research questions.

GIS: The Community

Many aspects of the behavior of disciplines are best understood from a sociological perspective: In many ways disciplines are like tribes, with traditions, loyalties, totems, icons, and symbols of membership. For debating purposes, GIS as the abstract area of interest may take human form in a caricature of the typical GIS specialist, a person obsessed with technology; tainted by association with Big Science, the military, and the security agencies; unethical; entrepreneurial; and politically conservative. Taylor (1990) refers to the "GISer," a term I believe I have never heard used by people in the GIS community to refer to each other—in fact I wonder whether any such term exists, and that in itself may be indicative of the loose cohesion of the field. From inside, an academic community provides stimulus and support, but from outside, it is easy to see a group as "other," especially when suspicions are fostered by an inability to understand the group's language. In return, the GIS community finds it easy to label more traditional fields as irrelevant and lacking in the impetus that derives from ties to activities outside the walls of academe.

In reality, the GIS community today is a loose consortium of interests, held together by a somewhat intangible enthusiasm for a poorly defined technology. It includes academics with solid research records, and others who one suspects could not survive without the protection of the group. It includes people with a deep understanding of the technology, and others who know it only as a black box. Like any other human group, it captures the rich diversity of the human condition.

Within U.S. geography, perhaps the most accessible way of identifying the GIS community is through the membership of the AAG (Association of American Geographers) GIS Specialty Group, now the largest in the association. Like specialty groups in general, it has a higher proportion of students than the association, and, despite its size, finds it difficult to foster a strong sense of belonging and a strong program of group activities. Perhaps this problem reflects the multidisciplinary nature of GIS—that loyalty to GIS and to geography is stronger than loyalty to a group of specialist geographers interested in GIS. I have seen the same pattern in other disciplines, where there is much interest in learning about GIS and in using it in the discipline, but a reluctance to develop a specialized group that may be in danger of becoming peripheral to the discipline as a whole.

GIS AS A TOOL FOR RESEARCH

Having offered three different views of GIS—the software, the set of research issues, and the stereotype GISer—I will now examine the role of GIS as a tool for research. The notion that one could automate the handling and analysis of spatial information—Dobson's "automated geography" (Dobson, 1983)—has intrigued geographers and others for many years. Physicists routinely use pattern recognition techniques to process the vast numbers of photographs produced by nuclear reaction experiments, and remote sensing specialists automate the interpretation of images, so why not extend these concepts to the more sophisticated and complex analysis and interpretation of geography and related disciplines?

GIS is now being used routinely by researchers in many disciplines. Although some of these disciplines, such as archaeology, geology, or transportation science, have traditionally employed a spatial perspective, in others GIS has generated a new interest in space and spatial thinking. Some of the more interesting applications of GIS in social science are emerging in history, sociology, criminology, and economics, all disciplines in which spatial thinking has played a very minor role in the past.

In practice, the part played by GIS in all these activities varies markedly, and only the software remains constant. At the most elementary level, space acts as little more than a convenient index, a means of arranging information in manageable form. Thus the archaeologist may map artifacts at a site simply because that is a convenient way to organize them. In this role, GIS acts as little more than a mapping system, allowing the user to manage data in an organized fashion, and to present them in convenient and readily understood ways.

At a somewhat more sophisticated level, GIS is used as a tool for preprocessing data prior to modeling or analysis. An environmental modeler will likely write his or her model in source code, typically FORTRAN or C, but may well maintain a GIS, linked to the modeling system, to preprocess data, and to analyze and present the model's results. This type of GIS use probably characterizes the majority of efforts in environmental simulation modeling at this time in disciplines like forestry, atmospheric science, or ecology; its state of development is extensively reviewed in Goodchild, Parks, and Steyaert (1993).

Interest in GIS in the emerging discipline of landscape ecology (Turner & Gardner, 1991) takes a rather different form. Recent research in biodiversity, gap analysis, and related areas has led to concern for simple geometric properties of ecological landscape, such as shape, and their role in determining habitat quality. GIS, with its emphasis on simple geometric analysis and spatial relationships, is an obvious toolbox for supporting such research.

In landscape architecture and related areas of resource management, the analysis functions of GIS perform a direct role in problem solving, modeling, and decision making. Here GIS is equivalent to a programming language, and the commands of GIS are in simple correspondence to the primitive operations needed by the scientist or manager. GIS has perhaps reached its highest level of analytic sophistication in these disciplines, where efforts have been made to codify the appropriate set of commands into simple spatial languages. Tomlin (1990) gives an excellent treatment of this field.

For many other scientists, GIS is a toolbox with useful commands, but some form of coupling must exist with other types of software in order to create a complete research environment. GIS typically do not contain statistical functions, or optimization routines, so it is common to find GIS coupled with the statistical packages: SAS, SPSS, S, and the like. Other specialized research may require the development of special modules, written in source languages like C, and coupled with the GIS. In these cases GIS performs the role of a general-purpose manager of spatial data.

This extensive adoption of GIS as a useful research tool has gone on with little regard for disciplinary boundaries, and despite the limited state of development of the technology. One now finds software like ARC/INFO installed widely on major research campuses—at Santa Barbara, we now have well over 100 licenses distributed across half a dozen departments. Geography has often been the initiator, but GIS is as likely to be found in biology, geology, anthropology, or any other discipline in which a spatial perspective is useful and insightful. However, GIS is most likely to be taught in one of the disciplines that contribute to geographic information science, particularly geography but also civil engineering, surveying, and geodesy.

Of all these disciplines, geography is clearly the one most able to close a critical gap in the use of GIS, the ability to combine an understanding of real geographic phenomena with the issues of their representation in a spatial database. Spatial representation is a strong part of a geographic education, and so too is a broadly based understanding of processes that affect the geographical landscape. Thus if the key issues of GIS are those of spatial representation in digital form, as I argued earlier in this chapter, then geography is the discipline most equipped to address them. It is the geomorphologist who is best able to choose the data model for representation of terrain in a GIS, not the computer scientist or the statistician, and it is the urban geographer who is best able to advise on how to represent the many facets of the urban environment in a GIS designed for urban planning. Outside geography, there seems to be widespread acceptance of this position, and a genuine willingness to hire geographers to provide the conceptual and intellectual framework for GIS.

GIS AND ETHICS

Many aspects of the GIS phenomenon continue to puzzle me. Why does the GIS community—defined, say, as the group of 3,000 or so people who attend one of the annual GIS/LIS (Land Information Systems) conferences—attract such a wide range of people, from computer hackers to map collectors? Maps are attractive because they are visual and they stimulate the imagination, and perhaps also because they present the world as simpler, more orderly, and less dynamic than it really is. Computers are attractive because they give power to their users, convey prestige and status, and behave in orderly ways. Is GIS attractive because these two sets of factors are somehow complementary? Or is it attractive because it allows people without training in cartography to make maps, people with little training in geography to analyze geographical distributions, and people without mathematical skills to model spatial phenomena?

Whatever the reasons, the strength of the GIS phenomenon is indisputable, and nowhere more so than in agencies of the federal government. In the past few years, the U.S. Environmental Protection Agency, the U.S. Forest Service, the Bureau of Land Management, the U.S. Geological Survey, and the Department of the Navy, among others, have all undertaken or initiated major procurements of GIS. In September 1992, the Department of the Navy announced the award of a contract valued at over $400 million to Intergraph Corporation for the supply of several thousand GIS workstations, with associated services, and a similar-sized procurement by the U.S. Forest Service was under way. Why would an agency like the Forest Service, charged with management of the nation's National Forests, undertake such a procurement and at the same time reduce staffing of backcountry stations, expenditures on fire protection, and other more traditional forms of resource management? Does society really want a Forest Service of GIS users at computer terminals rather than one of rangers on horseback?

It is easy to speculate on explanations, but it would be better to do so within the frameworks provided by the literature and major theories of social processes. GIS is now too widely adopted to be ignored;

perhaps it is the most significant event in spatial data handling since the invention of the map. We need to understand GIS's success, and the statements that it makes about the nature of society and its organizations. Is the Forest Service procurement part of an inevitable process in our litigious society toward management practices that are standardized and procedural, and therefore more open, and more defensible in court? Does it reflect a desire by management to control the actions and decisions of the organization, and an unwillingness to trust individuals to make decisions? Is a GIS user empowered by the technology, or demeaned by it? In many ways the GIS phenomenon surely mirrors patterns already evident in the adoption of other technologies in large organizations, but seldom have these been of such magnitude, and made at such cost. Although these issues clearly have little relevance to scientific users of GIS, they are important parts of the GIS research agenda for less technically minded geographers. Perhaps the recent literature on ethics and GIS cited earlier, and the other chapters of this book, will be the beginnings of a productive literature on GIS by geography's social theorists.

REFERENCES

Burke, L. (1993). *Environmental equity in Los Angeles* (Technical Rep. No. 93-6). Santa Barbara, CA: National Center for Geographic Information and Analysis.

Burrough, P. A. (1986). *Geographical information systems for land resources assessment.* Oxford: Clarendon Press.

Chrisman, N. R. (1988). The risks of software innovation: A case study of the Harvard Lab. *American Cartographer, 15*(3), 291–300.

Couclelis, H. (1991). Requirements for planning-relevant GIS. *Papers in Regional Science, 70*(1), 9–20.

Craig, W. J. (1989). URISA's research agenda and the NCGIA. *Journal of the Urban and Regional Information Systems Association, 1*(1), 7–16.

Dobson, J. E. (1983). Automated geography. *Professional Geographer, 2,* 135–143.

GIS World, Inc. (1991). *1991/92 International GIS Sourcebook.* Fort Collins, CO: Author.

Goodchild, M. F. (1988). Stepping over the line: Technological constraints and the new cartography. *American Cartographer, 15*(3), 311–319.

Goodchild, M. F. (1992a). Geographic data modeling. *Computers and Geosciences, 18*(4), 401–408.

Goodchild, M. F. (1992b). Geographical information science. *International Journal of Geographical Information Systems, 6*(1), 31–46.

Goodchild, M. F., Parks, B. O., & Steyaert, L. T. (Eds.). (1993). *Environmental modeling with GIS.* New York: Oxford University Press.

Heywood, I. (1990). Geographic information systems in the social sciences—introduction. *Environment and Planning A, 22*(7), 849–852.

Maguire, D. J. (1990). A research plan for GIS in the 1990s. In M. J. Foster & P. J. Shand (Eds.), *The Association for Geographic Information yearbook 1990* (pp. 267–277). London: Taylor and Francis.

Maguire, D. J., Goodchild, M. F., & Rhind, D. W. (Eds.). (1991). *Geographical information systems: Principles and applications.* London: Longman.

Monmonier, M. (1992). Summary graphics for integrated visualization in dynamic cartography. *Cartography and GIS, 19*(1), 23–36.

National Center for Geographic Information and Analysis. (1989). The research plan of the National Center for Geographic Information and Analysis. *International Journal of Geographical Information Systems, 3*(2), 117–136.

Openshaw, S. (1991). A view on the GIS crisis in geography, or, using GIS to put Humpty-Dumpty back together again. *Environment and Planning A, 23*(5), 621–628.

Openshaw, S. (1992). Further thoughts on geography and GIS—a reply. *Environment and Planning A, 24*(4), 463–466.

Peuquet, D. J. (1984). A conceptual framework and comparison of spatial data models. *Cartographica, 21*(4), 66–113.

Pickles, J. (1991). Geography, GIS, and the surveillant society. *Papers and Proceedings of Applied Geography Conferences, 14,* 80–91.

Samet, H. (1990). *The design and analysis of spatial data structures.* Reading, MA: Addison-Wesley.

Smith, N. (1992). History and philosophy of geography—real wars, theory wars. *Progress in Human Geography, 16*(2), 257–271.

Star, J. L., & Estes, J. E. (1990). *Geographic information systems: An introduction.* Englewood Cliffs, NJ: Prentice-Hall.

Taylor, P. J. (1990). GKS—A comment. *Political Geography Quarterly, 9*(3), 211–212.

Taylor, P. J., & Overton, M. (1991). Further thoughts on geography and GIS—A preemptive strike. *Environment and Planning A, 23*(8), 1087–1090.

Thompson, D., & Laurini, R. (1992). *Fundamentals of spatial information systems.* London: Academic Press.

Tomlin, C. D. (1990). *Geographic information systems and cartographic modeling.* Englewood Cliffs, NJ: Prentice-Hall.

Tomlinson, R. F. (1988). The impact of the transition from analogue to digital cartographic representation. *American Cartographer, 15*(3), 249–262.

Turner, M. L., & Gardner, R. H. (Eds.). (1991). *Quantitative methods in landscape ecology.* New York: Springer-Verlag.

CHAPTER 3

Geographic Information Systems and Geography

Peter J. Taylor
Ronald J. Johnston

The portrayal of how geography has developed must always be a reconstruction. (Olavi Grano, 1981, p. 17)

GIS AS AN OUTGROWTH OF THE "QUANTITATIVE REVOLUTION"

The recent phenomenal growth of interest in geographical information systems (GIS) within geography is not an autonomous process. Rather, it is part of the discipline's ongoing development and as such can be interpreted using the concepts devised in historiographic studies of geography. In fact, the rise of GIS fits very neatly into the contextual approach to the history of geography, which focuses on balancing the influence of "external" and "internal" factors on disciplinary change (Stoddart, 1981). Obviously GIS represents the interface between geography and an external technology, because developments in computers have been the key enabling factor that has made GIS possible. In this chapter we concentrate on the other half of the story, the much more contested question of how GIS emerged out of intellectual trends within geography.

The "quantitative revolution" provides a benchmark for considering contemporary geography (Taylor, 1991), and this is particularly the case for GIS. In the 1950s and 1960s a powerful intellectual movement developed within geography, one that succeeded in dominating the discipline by 1970. Sharing a severe disenchantment with regional geography and its emphasis on qualitative interpretations, a generation of geographers consciously set out to create a "new geography." Basically, the chief concern of these "revolutionaries" was to bring geography out of its "exceptionalist" isolation and to relocate it in the mainstream of trends in modern science. Geography was to be remade as a "science," thereby turning it into "quantitative geography" (Johnston, 1991).

Much subsequent discussion of this movement has been concerned with later intellectual reactions to what was to become known as positivism. The quantifiers were criticized from a range of contrary positions for their excessively narrow interpretation of what constitutes science. In this process the quantitative revolution was reconstructed as a unitary monolith and any diversity associated with its theoriticians tended to be written out of the story. Yet the changes that occurred in the 1950s and 1960s were anything but unitary. A simple comparison of the key texts of the period confirms this point: It would be hard to find a more disparate set of books than William Bunge's (1962) *Theoretical Geography,* Stan Gregory's (1963) *Statistical Methods and the Geographer,* Peter Haggett's (1965) *Locational Analysis in Human Geography,* and David Harvey's (1969) *Explanation in Geography.* (The first deals with spatial science, the second with statistical techniques, the third with social science locational models, and the fourth with the nature of models and theories in science.) Contemporaries were very much aware of and concerned about these differences and debated what their revolution should be called—three early popular labels were "conceptual," "model-based," and "statistical"—before the label "quantitative" was generally adopted, to which Burton (1963) added "theoretical geography," thereby making a formal link between rigorous methods and a "scientific approach." The variety of these key texts and labels reflected crucial tensions within the movement for change, as we would expect to find in any "revolutionary situation." The later emergence of GIS is related to two such tensions, that between deductive and inductive "science" and that between "pure" and "applied" geography.

The first tension is illustrated by the contrast between Gregory's (1963) argument that geographers should use rigorous tools in their empirical analysis and Bunge's (1962) exploration of spatial theory in geography. Gregory's revolution was envisioned as a retooling, whereas Bunge represented a more radical majority position that called for much more than a technical revolution (as Burton, 1963, indicated in his classic paper; see also Davies, 1972). New statistical and mathematical methods were increasingly seen as a necessary but not sufficient condition for transforming geography. In the 1970s this way of thinking led to widespread use of systems analysis in geography. In systems modeling, urban land use and transport were the growth areas (Wilson, Rees, & Leigh, 1977); in applications of systems theory, social–environmental interactions came under scrutiny (Chapman, 1977; Bennett & Chorley, 1978). Their commonality was that systems were specified before any empirical analysis: The quantifiers were distancing themselves from potential criticisms of "empiricism." In the 1980s GIS has shown no such qualms; it is a collection of quantitative tools for data analysis. In this context, its proponents represent a throwback to the early "quantitative techniques only" approach to changing geography, eschewing the mainstream concern for models and theory. They are strictly empirical quantitative geographers.

The second tension within the quantitative revolution was between "pure" and "applied" studies. The whole argument for a revolution was premised upon the failure of geography to become respectable science. If geography did not constitute a rigorous body of knowledge, its position within universities would be indefensible. In the scientific pecking order "pure" theory, and a form of theory that was as mathematical as possible, was identified as having the highest intellectual status, and hence this defined the path for geography. The development of theory became the central task of the revolutionaries. Haggett's (1965) widely used text was a review of models and theories for human geographers and Harvey's (1969) text provided the scientific guidelines for such theoretical development. Christaller's central place lattices, Hortonian networks, Stewart's population potentials, Burgess's urban zones, Weber's locational triangles, von Thunen's "isolated state," and more, were all moved from either the periphery or outside the discipline to its very core to constitute the new geography (Haggett, 1965; Pooler, 1977). The valorization of "pure" geography was seen to be essential if geography was to survive in academia.

Of course, geographers did not abandon applied studies. Central place theory could be used in settlement planning, for example, but this was never seen as the theory's ultimate purpose. It could and should be applied for practical purposes, but the theory was developed for its own sake to aid in understanding the spatial patterns that are geography's subject matter. To the revolutionaries, applications were a secondary activity, so applied geography commanded none of the status of "pure" geography.

All this changed after 1970 as the world economy moved into a downswing and criteria for disciplinary preservation in academia were changed accordingly (Taylor, 1985). In these new trying times disciplines justified themselves by being useful. Concern for spatial theory began to be seen as an indulgence that could no longer be afforded (Openshaw, 1989). "Relevance" was suddenly the buzzword in geography, and in a rapid reversal applied geography became extremely fashionable in the 1970s (Johnston, 1991). The systems geography referred to above reflected this process. While developing their systems models and theories, geographers were concerned to apply their analyses, whether in predicting traffic flows in city systems or in evaluating environmental impacts in river basin systems. For some, "applied geography" even reached the status of a new paradigm for geography (Frazier, 1978). This concern for application has continued into the 1980s and is central to the success of GIS.

In this second tension GIS selectively draws from these later trends in quantitative geography. The proponents of GIS combine the early technical concerns of quantifiers with the later social, economic, and political concerns of those who advocate applied geography. Hence, although they derive their positions from trends set in motion by the quantitative revolution, as a major tendency in contemporary geography GIS practitioners constitute a new grouping of ideas. They are strictly applied quantitative geographers.

APPLIED GEOGRAPHIES

Most of the calls for more applied geography have been narrowly based within the discipline, resting firmly on the positivist foundations established by the promoters of the "quantitative revolution." As far back as 1979, Anne Buttimer sought to focus geographers' attention on an applied geography in which the professionals acted not as social engineers, who manipulated people and forced change, but rather as provocateurs, who helped individuals to understand both themselves and others and so were better able to chart their own futures. The message appears largely to have fallen on deaf ears: A decade later, Pacione

(1990) presented his vision of an "applied urban geography" in which the focus was problem-solving (not resolving) scientific activity (for a critique, see Johnston, 1990).

It is into the framework exemplified by Pacione's paper that GIS is being fitted by its proponents: This framework offers ways of analyzing data and producing "solutions" which, because "better" than the present situation (and "better" usually means more efficient in either cost-minimization or utility-maximization terms, or both), are to be used in beneficent sociospatial engineering. And yet that framework represents only a part (almost certainly a minority part) of contemporary geography, and people seeking to meet the challenges of a rapidly changing world are as likely to base their arguments on other philosophies as they are to embrace the empiricist-cum-positivist stance taken, at least implicitly, by most GIS advocates.

The myopia of the "quantitative revolution" was clearly brought to geographers' attention by Derek Gregory's (1978) seminal book, *Ideology, Science and Human Geography.* Drawing on Habermas (1972), he identified three types of science, from which three very separate applied programs have been derived (Johnston, 1992).

1. Empiricist–positivist geography and technical control equate prediction with explanation. Desired ends can be defined and then attained, it is argued, because knowing the preconditions of events means that the events can either be avoided (by removing such necessary conditions) or promoted (by engineering their precursors). Whereas some geographers believe that only the physical world can be treated in such a way, others argue that both individuals and the societies they create are open to similar "science-based" manipulation. This is the usage of GIS promoted by Openshaw (1989) in his empiricist "value-added-to-data" presentation of a geography focusing on "the transformation of knowledge into information which can be exchanged, owned, manipulated and traded" (p. 81).

2. Humanistic science and mutual understanding promote appreciation of events involving humans by uncovering and communicating the thoughts behind action. The only explanation sought is that provided by the actors themselves, either directly or indirectly (see Sayer, 1992). Geographical analysts communicate their understandings of those thoughts and their outcomes with the goal not of manipulating people but rather of enhancing both self-awareness and mutual awareness. On the basis of that enhanced appreciation in which

the analyst, whose own position should be clearly stated, cannot act in a value-free manner, individuals can promote their own futures.

3. Realist science and emancipation seek to enhance and, where necessary, correct the possibly flawed interpretations that individuals have of their personal situations, and that may be revealed to them by humanistic science. People's interpretations of how the physical and social worlds operate may be sufficient for them to survive, but if those understandings are imperfect then their potential to achieve desired change is reduced. Realist science seeks to avoid such situations by eliminating individuals' distorted views (many of which they will have developed through processes of socialization in a society that either deliberately or unintentionally seeks to sow false understandings). By identifying the true causes of events, geographers operating as scientific realists give people a much greater ability to control their own futures.

The vigorous promoters of GIS in recent years have clearly set their offerings strictly within the context of the first of these three types of knowledge-constitutive interest, and their corresponding forms of science. They have thus been advancing a particular and partial view of geography, as Openshaw's (1989, 1991) pronouncements exemplify, alongside the less strident claims of Bennett (1989), Rhind (1989), and others. GIS allows spatially referenced data to be used in ways heretofore impossible, and thus provides spatial scientists with opportunities to contribute to the spatial restructuring of societies, which they may do serendipitously—Openshaw (1989) merely sees GIS geographers as the sellers of data to would-be users, for example, whereas Bennett (1989) sees them contributing to the decline of welfarism and the growth of a market-based economy (see Johnston, 1993).

In this context, it is important to recognize the criticisms that have been mounted against the spatial science paradigm, many of them more than a decade old, and that have led to spatial analysis in British human geography being identified as a "twenty-year diversion" (Johnston, 1985). In brief, those critiques argued that the empirical generalizations that the positivist-inclined human geographers were producing were descriptions without any explanatory base. This case was made especially strongly by Sack (1972, 1973, 1974) in a series of papers that contested Bunge's (1966) and Harvey's (1969) claims regarding geometry as the spatial language that geographers should adopt. Geometric language can describe a spatial arrangement (as ably illustrated in Haggett's [1965] seminal text), but it cannot explain: Knowing that the impact of an axe on wood will split it does not tell you why the axe is being directed at the wood in the first place. The original "quantifiers" attempted to counter this argument by developing deductive theory but, as we have noted above, it is just this aspect of quantitative geography that has been severely castigated by GIS proselytizers (Openshaw, 1989; see also Batty, 1989).

GIS development has been set within the context of a relatively myopic view of geography, resulting in an applied agenda that has many inherent flaws. This is not to deny its utility in many circumstances, as we argue below, but it does indicate a necessity to appreciate where it has come from in order to understand where it is going and what its proponents are trying to do. Many (most?) geographers want to understand the worlds in which they live, in order that they can help people to create better worlds: That goal involves much more than identifying empirical regularities in spatial forms and movement patterns.

THE DISUTILITY OF GIS

Although computers were important during the quantitative revolution, they set severe limits on the size of data sets that could be analyzed. By 1970 multivariate analyses of U.S. Bureau of the Census variables over urban census districts had become commonplace as larger computers and associated software became generally available. Although immensely impressive to quantifiers at the time, this computational capacity still kept spatial analysis within conventional statistical bounds. GIS is all about breaking those bounds with the increased data manipulation capacity of later generation computers. Technical breakthroughs have led to wider horizons in spatial analysis never even dreamed of in 1970. Very large scale data sets are now available for manipulation. A whole new world of data analysis has been opened up.

A basic problem with this development occurs when it is used to propose a new geography that is data-led (Openshaw, 1991). The determining factor in such a construction is the availability of data. This means that anybody concerned with the nature of geography in the light of these trends must confront the nature of the data upon which it is to be built. One of the most disappointing features of GIS is the failure to address this question seriously. Data are usually treated unproblematically except for technical concerns about errors. But data are much more than technical compilations. Every data set represents a myriad of social relations. The basic relation is that a person (A) decides to collect information on another person or a thing (B). Thus there is an implicit power relation; in general, the more powerful do the finding out about the less powerful. But they do not find out everything, since A selects what it is about B that is useful to collect. Finally, A as owner of the information about B can decide what to do with it. When stated baldly like this, data are clearly anything but unproblematic from a social viewpoint (a point intriguingly recognized in Openshaw & Goddard, 1987; see also Beaumont, 1991).

Given the above, it should come as no surprise to our readers to discover that *statistics* and *state* come from the same root. The state represents a concentration of formal power that both facilitates and relies upon the collection of information. Modern statistics began to be collected with the emergence of the modern state. Today vast amounts of data emanate from the many state apparatuses. These have been by far the major source of data for GIS practitioners in their production of new geographies. With this information they are thereby creating the state's geography, so that the state, through its data collection decisions, determines the agenda of this geography, that is, what is included and what is not. This would be fine if it were one of several geographies being produced but since the state's data dominate all data that are publicly available, a GIS geography can thus be condemned as a handmaiden of the state.

This situation would not be a problem if we all subscribed to a benign theory of the state. If the state truly represents the public interest, then a geography steered by state data collection would similarly be in the public interest. But, this proposition is contestable in liberal democratic states and therefore should be debated within GIS-constructed geography. The problem with GIS geography is that it is committed to one side of the debate before discussion begins. This problem is a much greater one beyond the minority of (mainly liberal-democratic) states located in the core of the world economy. Elsewhere the state is anything but benign; collecting information about individuals where death squads operate is not to be encouraged. Fortunately, such states are much less integral to world society and their capacity to collect information is also much less. This leads to a second problem for a GIS geography.

In dealing with core states, we noted that the geography agenda is distorted by being data-led. A GIS geography implies a neglect of themes not included in the data. Such a position has been advocated within GIS (Openshaw, 1991). When we move to the global scale, this has serious implications for world coverage by geographers, because it produces what has been termed ''the first law of geographical information'': The poorer the country, the less and the worse the data (Taylor & Overton, 1991). Hence a data-led GIS geography would neglect most of the world. Analysis dependent upon rich data sources produces a rich countries' geography (see also Taylor, 1992).

Of course the meaning of such neglect is much more than geographical omission. GIS geography as a data-led project is inherently empiricist. This does not mean that it is ''theory-free'' in any sense, since all analysis is built upon social theory—some explicit, some implicit. A GIS geography that does not consider its underlying theory is avoiding its responsibility and investing it in the collectors of its data, usually the state. Such ''empiricism'' produces an inherently conservative geography. The theory upon which analysis is built will be descriptive of the status quo and treat it as a taken-for-granted world. Alternative worlds—for which there are no data—are ruled out of court in this empiricist development of geography.

There is another important sense in which GIS geography can be inherently conservative in its prescriptions. As we have seen, GIS does not operate through connections derived from systems thinking; rather, it treats data as discrete entities for which empirical relations can be sought. But the data are outcomes of processes and mechanisms that can only be understood through some knowledge of how the data fit together as social organization (which means a causal and not a mechanistic appreciation of process; see Hay & Johnston, 1983). Some theory, perhaps of the systems variety disregarded in GIS, is required to make sense of the pattern of outcomes recorded. As was pointed out at the very beginning of the quantitative revolution (Blaut, 1962), the basic problem of pattern analysis is that different processes may produce the same pattern and the same pattern may be produced by different processes. No amount of sophisticated analysis of patterns in GIS can solve this problem. Instead, we need a theory that distinguishes between contingent and necessary relations in the outcomes represented by the data (Sayer, 1992). Ontologically shallow GIS analyses lack the theory to deal adequately with this crucial problem.

Failure to confront this problem in applied geography produces comparative geographical analyses that lead to simplistic policy prescriptions. For example, many studies involve ranking places on one or more criteria, and allocating policy benefits accordingly. At its crudest this applied geography merely provides a list of winners and losers with no understanding of why the differences occur. The implication is that every place being ranked can be equally successful: All that is required is that the "losers" emulate the "winners." The implicit philosophy (termed "the error of developmentalism" by Taylor, 1989) is that if the latter places can do it, why not the former? This position would be reasonable if we lived in a functionally undifferentiated world. But we do not. Our world is a highly connected one of interrelated and overlapping systems. We do not have to hypothesize a zero-sum game to see that success in one place precludes to some degree its repetition elsewhere (Johnston, 1986a). Simply not everywhere can be the same, and winning is not an autonomous process: To think otherwise is to commit, in systems terms, the fallacy of composition (Sayer, 1992).

The basic point of this section can be summarized by the question "Whose geography?" A data-led geography is a geography of the data gatherers, the powerful. But there are other geographies. Can GIS contribute to them?

THE UTILITY OF GIS

In putting GIS in its intellectual context within the history of geography, and in pointing to the potential dangers in adoption of the data-led and market-based orientation to its use which some at least have propounded, it has not been our goal to decry GIS entirely, nor to argue that it lacks utility within either the practice of geography or the contributions that applied geography can make to the society within which it operates. In this, as in other aspects of the link between GIS and the "quantitative revolution" which spawned it, we are arguing for its proper use and against the inflated and intellectually naive claims that some have made on its behalf.

Unfortunately, some critics of quantitative geography attacked what was done not only on the well-founded grounds set out in the critiques discussed above, but also on the basis that any quantitative work (beyond a bare minimum) must of necessity be philosophically flawed. This was the position taken by Sayer (1984) in the first edition of his influential *Method in Social Science* and continued—though more mutedly—by him in the second edition (Sayer, 1992), despite arguments to the contrary by, inter alia, Johnston (1986b) and Pratt (1989). Thus Sayer correctly refers to mathematics as an "acausal language" and argues that repeated observation of a regularity in no way constitutes an explanation for it. But he then assumes that (virtually) all quantitative applications necessarily seek to provide both explanations and predictions. If this were a valid claim, then it could be extended to GIS as well. But it is not a valid claim. Quantitative methods can properly be applied in a descriptive mode, as a way of untangling a complex of material without either implying causation or suggesting that the same outcome might recur. (However, the untangling may be structured to evaluate a hypothesis that implies causation. The statistical work may produce findings consistent with expectations, whose value depends on the force of the causal argument that led to the hypothesis generation and not on the "statistical significance level" of the analyses themselves.) Sayer may well be right in his claim that "Often the price of achieving mathematical order and rigor is conceptual sloppiness produced by disregard of the nature of the object being modeled" (p. 201), but this is a positive and not a normative claim: Just because practitioners are sloppy does not mean that they *have to be* sloppy! Sayer is centrally concerned with the difference between "vary with" and "causally determine" (terms that many researchers appear to treat as synonyms). That difference, he claims, "marks a divide between radically different kinds of research with very different chances of providing illuminating answers" (p. 203). We concur, but we differ with Sayer by our willingness to accept the possibility that quantitative procedures, and hence GIS, can provide illuminating material which poses important questions even if it cannot produce substantial answers to the question "Why?"

The position that we take here was outlined more than a decade ago by one of us in a discussion of the use of factor analysis in geographical research (Taylor, 1981, p. 251). Since it is only a technique it is hardly fair to criticize it for how we have used it. Like all techniques, it does some jobs reasonably well and other jobs poorly. Factor analysis is a relatively sophisticated "measurement technique" and, as with similar developments in psychometrics, biometrics, and econometrics, much of the opprobrium it received in the late 1970s reflected its use, abuse, and misuse by relatively conservative scholars: "It is merely a technique, a tool, which has proved useful to conservatives in the past, but that is no reason for non conservatives to discard it from their battery of techniques to criticize the present world'' (Taylor, 1981, p. 257).

In the original text, this argument was illustrated by material drawn from an investigation of the changing geography of electoral behavior in the United States, a study clearly placed within a realist theory derived from Wallerstein (Archer & Taylor, 1981; see also Taylor, 1988). The claim for this analysis was a modest one: a presentation of ''a relatively simple measurement exercise in which the major structural constraints on American elections have been delineated. This is a job that factor analysis does well and we can ask little more of any tool than that it does its job well'' (Taylor, 1981, p. 265). As with factor analysis, so with GIS.

If GIS is a tool, what jobs does it do well? GIS is essentially a technology that can be used for ''handling, processing and analyzing geographic data'' (Goodchild, 1993). Its advantages over previous technology, both hardware and software, largely reflect the massive increases in cheap computing power during the last decade. From these, it is possible to assemble large data sets referring to point, line, and areal distributions (with the assembly itself perhaps automated, as in the applications of remote sensing data collection and assembly procedures), to integrate them, and then to process them (which may include linking them in various ways, both to add value to them in the manner advanced by Openshaw, 1989, and to use them to test for spatial associations that may be of value in the evaluation of putative explanatory accounts).

The power of such technology should not be underestimated; nor should GIS, as quantitative methods were earlier, be either overpromoted or attacked on spurious grounds. Openshaw (1991) argues, rightly in our view, that the new technologies epitomized by GIS are providing tools for geographers (and others) to use on geographical information. What they are used for and how to make best use of them within geography depends on the attitudes and mindset of their users and what they want to do with them. In fact, all that is really changing is the manner by which geographers can perform some of their more explicitly geographical works and the appearance of an information framework within which all geographers should be able to work. But this basic utility argument is prejudiced by simultaneously promoting a disutility argument related to our previous discussion. GIS will, according to Openshaw (p. 622):

> have a far-reaching, long-term, and fundamental impact on the nature of geography itself. GIS . . . can be regarded as offering the prospect of reversing [geography's] disciplinary fissioning process and replacing it by a fusion; a drawing together of virtually all the subdisciplinary products with their multitude of conflicting paradigms created over the last thirty years, within a single (philosophy-free or philosophy-invariant or even philosophy-ignorant) integrating framework.

Such inflated claims suggest an immense näiveté in failing to understand the various forms of geography that have developed in recent decades and in equating information with knowledge (Taylor & Overton, 1991). Such claims are perhaps typical of propagandists, within and outside geography, but do a good cause no good at all. They reflect acceptance of the 1980s propaganda of the "New Right" regarding the superiority of the market over all other forms of distributing the means to economic and social well-being (Bennett, 1989, Johnston, 1992).

Geographers of all persuasions and all applied agenda are interested in making full use of available information and in doing so as efficiently and as effectively as possible. GIS offers them much. They are now able to collate large data sets in ways previously considered impossible, and to interrelate those datasets, again in ways that offer power and sophistication in ways undreamed of even a decade ago. They can portray the world in a complexity and detail that their predecessors could hardly have imagined. They can explore patterns, they can evaluate the importance of such constraints to geographical representation as spatial autocorrelation and the modifiable areal unit problem, and they can portray the world in new formats. Openshaw himself has taken the lead in many of these applications, as exemplified in his work on the etiology of disease (Openshaw, Charlton, Craft, & Birch, 1988) and the potential spuriousness of correlations based on areally aggregated data (Openshaw, 1977). Why claim more? And although GIS specialists are mainly publishing articles on the development of their field (as illustrated by the large number of papers in the *International Journal of Geographical Information Systems* on data structures and database management—shades of early papers in the "quantitative revolution")—nevertheless, they are addressing the use of GIS in tackling substantive research questions (as in Carver, 1991).

Many research projects, like projects in everyday life, are impelled by images of the world. We receive and/or develop an image, often through either or both of selective filters and rose-colored spectacles (Lowenthal, 1961), and use it as a lead-in to the questions that we pose and the direction of our search for the answers. GIS can provide us with such images, and potentially will produce such a wealth of images that we may not be able to cope with them all—so that our selectivity will be driven by other images. Like the quantitative methods that preceded their development, geographers will use those image-development and image-enhancement tools to promote those goals of their discipline that are central to their own research and personal agenda. It may well be that such promotion will be concentrated within one of the three types of applied geography identified above, but there is no reason for a ghettoization of GIS in the technical control program and no reason against their full use as tools within the other two programs.

A spade, according to a 1992 advertisement on British television, can be presented either as a tool to be used in the oppression of the working class or as a means of liberating the potential of would-be innovators. GIS is seen by some critics as falling into the former category, and is overpresented by some of its champions as the widely sought panacea to disciplinary ills which the latter might provide. As so often happens, its real value lays in between, as an efficient tool for manipulating information, no more, no less. Unfortunately, as its intellectual context clearly shows, too often too much has been claimed for an innovation by some and too much opprobrium has been heaped on it by others. A spade, is a spade, is a spade. . . .

REFERENCES

Archer, J. C., & Taylor, P. J. (1981). *Section and party.* Chichester, England: Wiley.

Batty, M. (1989). Urban modelling and planning: Reflections, retrodictions and prescriptions. In B. Macmillan (Ed.), *Remodelling geography* (pp. 147–169). Oxford: Basil Blackwell.

Beaumont, J. R. (1991). The value of information: A personal commentary with regard to government databases. *Environment and Planning A, 24,* 171–177.

Bennett, R. J. (1989). Whither models and geography in a post-welfarist world? In B. Macmillan (Ed.), *Remodelling geography* (pp. 273–290). Oxford: Basil Blackwell.

Bennett, R. J., & Chorley, R. J. (1978). *Environmental systems: Philosophy, analysis and control.* London: Methuen.

Blaut, J. M. (1962). Object and relationship. *Professional Geographer, 14*(1), 1–7.
Bunge, W. (1962). *Theoretical geography.* Lund, Sweden: C. W. K. Gleerup.
Bunge, W. (1966). *Theoretical geography* (2nd ed.). Lund, Sweden: C. W. K. Gleerup.
Burton, I. (1963). The quantitative revolution and theoretical geography. *Canadian Geographer, 7,* 151–162.
Buttimer, A. (1979). Erewhon or nowhere land. In S. Gale & G. Olsson (Eds.), *Philosophy in geography* (pp. 9–38). Dordrecht, The Netherlands: D. Reidel.
Carver, S. J. (1991). Integrating multi-criteria evaluation within geographical information systems. *International Journal of Geographical Information Systems, 5,* 321–340.
Chapman, G. P. (1977). *Human and environmental systems: A geographer's appraisal.* London: Academic Press.
Davies, W. K. D. (Ed.). (1972). *The conceptual revolution in geography.* London: University of London Press.
Frazier, J. W. (1978). On the emergence of an applied geography. *Professional Geographer, 30,* 233–237.
Goodchild, M. F. (1993). Geographical information systems. In R. J. Johnston, D. Gregory, & D. M. Smith (Eds.), *The dictionary of human geography* (3rd ed., pp. 219–220). Oxford: Basil Blackwell.
Grano, O. (1981). External influence and internal change in the development of geography. In D. R. Stoddart (Ed.), *Geography, ideology and social concern* (pp. 17–36). Oxford: Basil Blackwell.
Gregory, D. (1978). *Ideology, science and human geography.* London: Hutchinson.
Gregory, S. (1963). *Statistical methods and the geographer.* London: Longman.
Habermas, J. (1972). *Knowledge and human interests.* London: Heinemann.
Haggett, P. (1965). *Locational analysis in human geography.* London: Edward Arnold.
Harvey, D. (1969). *Explanation in geography.* London: Edward Arnold.
Hay, A. M., & Johnston, R. J. (1983). The study of process in quantitative human geography. *L'Espace Geographique, 12,* 69–76.
Johnston, R. J. (1985). Spatial analysis in British human geography: A twenty-year diversion? *L'Espace Geographique, 14,* 29–32.
Johnston, R. J. (1986a). Understanding and solving American urban problems: Geographical contributions? *Professional Geographer, 38,* 229–233.
Johnston, R. J. (1986b). *On human geography.* Oxford: Basil Blackwell.
Johnston, R. J. (1990). Some misconceptions about conceptual issues. *Tijdschrift voor Economische en Sociale Geografie, 81,* 14–18.
Johnston, R. J. (1991). *Geography and geographers: Anglo-American human geography since 1945* (4th ed.). London: Edward Arnold.
Johnston, R. J. (1992a). Meet the challenge: Make the change. In R. J. Johnston (Ed.), *The challenge for geography—A changing world: A changing discipline* (pp. 151–180). Oxford: Basil Blackwell.
Johnston, R. J. (1993). The rise and decline of the corporate-welfare state. In

P. J. Taylor (Ed.), *The political geography of the twentieth century* (pp. 115–170). London: Belhaven Press.

Lowenthal, D. (1961). Geography, experience and imagination: Towards a geographical epistemology. *Annals of the Association of American Geographers, 51,* 241–260.

Openshaw, S. (1977). A geographical solution to scale and aggregation problems in region-building, partitioning, and spatial modelling. *Transactions of the Institute of British Geographers, N.S., 2,* 359–372.

Openshaw, S. (1989). Computer modelling in human geography. In B. Macmillan (Ed.), *Remodelling geography* (pp. 70–88). Oxford: Basil Blackwell.

Openshaw, S. (1991). A view on the GIS crisis in geography, or, using GIS to put Humpty Dumpty back together again. *Environment and Planning A, 23,* 621–628.

Openshaw, S., Charlton, M., Craft, A. W., & Birch, J. (1988, February 6). Investigation of leukaemia clusters by use of a geographical analysis machine. *The Lancet, 155,* 272–273.

Openshaw, S., & Goddard, J. B. (1987). Some implications of the commodification of information and the emerging information economy for applied geographical analysis in the United Kingdom. *Environment and Planning A, 19,* 1423–1440.

Pacione, M. (1990). Conceptual issues in applied urban geography. *Tijdschrift voor Economische en Sociale Geografie, 81,* 3–13.

Pooler, J. A. (1977). The origins of the spatial tradition in geography: An interpretation. *Ontario Geography, 11,* 56–83.

Pratt, G. (1989). Quantitative techniques and humanisitic-historical materialist perspectives. In A. Kobayashi & S. Mackenzie (Eds.), *Remaking human geography* (pp. 101–115). Boston: Unwin Hyman.

Rhind, D. W. (1989). Computing, academic geography and the world outside. In B. Macmillan (Ed.), *Remodelling geography* (pp. 177–190). Oxford: Basil Blackwell.

Sack, R. D. (1972). Geography, geometry and explanation. *Annals of the Association of American Geographer, 62,* 61–78.

Sack, R. D. (1973). A concept of physical space in geography. *Geographical Analysis, 5,* 16–34.

Sack, R. D. (1974). The spatial separatist theme in geography. *Economic Geography, 50,* 1–19.

Sayer, A. (1984). *Method in social science: A realist approach.* London: Hutchinson.

Sayer, A. (1992). *Method in social science: A realist approach* (2nd ed.). London: Routledge.

Stoddart, D. R. (Ed.). (1981). *Geography, ideology and social concern.* Oxford: Basil Blackwell.

Taylor, P. J. (1981). Factor analysis in geographical research. In R. J. Bennett (Ed.), *European progress in spatial analysis* (pp. 251–267). London: Pion.

Taylor, P. J. (1985). The value of a geographical perspective. In R. J. Johnston (Ed.), *The future of geography* (pp. 92–110). London: Methuen.

Taylor, P. J. (1988). History's dialogue: An exemplification from political geography. *Progress in Human Geography, 12,* 1–14.

Taylor, P. J. (1989). The error of developmentalism in human geography. In D. Gregory & R. Walford (Eds.), *Horizons in human geography* (pp. 303–319). London: Macmillan.

Taylor, P. J. (1991). A future for geography. *Terra, 103,* 21–31.

Taylor, P. J. (1992). Full circle, or new meaning for the global? In R. J. Johnston (Ed.), *The challenge for geography—a changing world: A changing discipline* (pp. 181–197). Oxford: Basil Blackwell.

Taylor, P. J., & Overton, M. (1991). Further thoughts on geography and GIS. *Environment and Planning A, 23,* 1087–1090.

Wilson, A. G., Rees, P. H., & Leigh, C. M. (1977). *Models of cities and regions.* Chichester, England: Wiley.

CHAPTER 4

Geographic Information Systems and the Inevitability of Ethical Inconsistency

Michael R. Curry

Geographers have lately engaged in much discussion of geographic information systems (GIS), but have devoted little attention to the ethical implications of their construction and use. It might be argued that there is an obvious reason for this failure: GIS is not completely new, but rather are merely continuations of already existing practices and forms of inquiry within the discipline, and so one need only look to those predecessors for guidance. But there are two problems with that response. Geography has almost no history regarding discussion of matters of ethics[1] and when one turns specifically to GIS, one is faced not with a single source of authority, but with multiple possible sources. There are literatures about ethics and science, the professions, technology, computers, business, and government; there is even a newly emerging literature on the ethics of representation.[2] Each of these literatures approaches matters from a slightly different perspective, and for that reason the student of the relationship between ethics and GIS can end up confused about where to turn.

In fact, I would argue that while one may very well profit from a consideration of these literatures, an understanding of the ethical issues that attend the construction and use of GIS needs to see those systems—because they combine in new ways issues raised in all these literatures—as having their own existence sui generis. It is necessary, that is, to construct an analysis that can see the ethical problems that GIS creates as a result of its own, specific nature. Here a truly adequate account of the ethical status of GIS needs to move beyond a consideration of the actual practice of automated cartography and GIS, and to consider the relationship of that practice to the contexts within which GIS operates: the field of geography, the broader science establishment, the information marketplace, and various levels of government.

This is a large undertaking, and my intention here is not to complete the task, but merely to make a start. I begin by laying out what I take to be the typical way of talking about the relationship between GIS and ethics. This way of talking involves a series of assumptions about the activities of research and communication, as well as assumptions about the social practices involved in the maintenance and use of GIS. I shall then explore the nature of assumptions about ethics implicit there. As I have argued elsewhere (Curry, 1991d), these strains are expressive of a dissolution of the belief in the existence and efficacy of a set of functional bonds that hold together research, communication, and the social systems of science and of geography (Merton, 1973). Here I shall argue that in the case of GIS matters have gone further; fundamental features of those systems require that those who create and use them sustain disparate and contradictory sets of ethical discourses. If these contradictions are not unique to those who use GIS, they *are* especially pronounced there, and they lead to what I have termed the "inevitability of ethical inconsistency."

I draw four conclusions from an analysis of the relationship between GIS and ethical issues. First, and most simply, the creation and maintenance of GIS involve ethical inconsistencies. Second, these inconsistencies are not merely contingent features of current practice; rather, they are necessary features of the systems, built into them through the intersection of the technological and the social. Third, this intersection is expressed in the creation of sets of patterns of practical knowledge, where the explicit goals of one set may—and in fact do—conflict with the implicit goals of another; this is, in fact, what constitutes the contradictions of which I have spoken. Fourth, this shows the necessity of seeing a "realm," like GIS, as constituted of sets of interlocking and overlapping patterns of actions. Indeed, the popular alternative, to visualize academic disciplines or subdisciplines in terms of trees or other spatial images, tends in the end to obscure the perception of the very issues of which I have written.

GIS IN THE VERNACULAR VIEW

The normal view of the relationship between GIS and ethics is, I think, something like the following. There is a system of geographical knowledge and practice, and GIS forms a part of it. As with other parts of geography, in GIS matters proceed in a stepwise fashion. The geographer begins by obtaining data. Exactly how those data are obtained is not terribly important, although one hopes that they are reasonably accurate. Once accumulated, the data are incorporated into a system; in a sense, they now cease to be "brute facts," and instead become "information." The process of translation is carried on by the geographer, who analyzes the data, performs transformations, and ultimately represents the data in visual output. The output then goes to the end user, who puts the data to use (see, e.g., Burrough, 1986; National Center for Geographic Information and Analysis, 1989; Star & Estes, 1990).

Although I have mentioned a number of steps, all these steps may involve no more than a single individual, as perhaps when students are asked to collect data, write software, analyze the data, and create output. Of course, more typically these steps involve a range of actors, and these actors seem increasingly to be separated by vast amounts of physical, if not electronic, space. In the normal situation the two ends of the process—defining the problem and gathering the data at one end, using or applying the output at the other—are actually external to the process, and it is the middle to which one devotes one's attention.

When the systems are characterized in those terms, the ethical responsibilities of the creator or user of GIS or of automated cartography are reasonably straightforward. The practitioner needs to exercise "due diligence" or "reasonable care," to use two typical terms, to see that the data are reasonably coherent. The analysis of those data should meet similar standards. And the output ought to be, to put things in more traditional terms, a "good map," one that clearly communicates information in ways that are not biased, or muddled, or ideologically twisted. Having attended to these matters, the practitioner can feel confident that he or she has acted responsibly. Actually, the ideal here is to build as much as possible of these standards of responsibility into the system itself; when one creates a "turnkey" system, the entire middle of the process, the elements of analysis and production, is automated and routinized, and so in a sense are the standards of responsibility.

Just by simplifying it, I have caricatured the process. For one thing, most of us continue to worry about our products, whether maps, articles, or children, even after they are out the door. But at the same time, I think that this description is reasonably accurate. It seems to apply not just to GIS, but also to academic work more generally. Whether we are actively involved in creating and maintaining GIS or not, we amass data, analyze the data, and produce results. Just as in the case of GIS, we feel a responsibility to do good work. Indeed, one of the features of science that many would say makes it distinctive is that it is a social process that ensures that people will treat their work with just that responsibility. More important, this description actually applies not just to GIS and science, but also to work more generally. In our society, at least, most people feel enjoined to do a reasonably good job or, in other words, to act responsibly (see Zuckerman, 1984; Schmaus, 1983).

It is this fact—that the description applies not only to GIS, but also to people's activities in the workplace more generally—that I think makes it clear that this is just not an adequate account of the ethical content of the practice of creating and using GIS. Indeed, to believe that it is is to dismiss the claims of advocates of GIS about the importance of their project. So the question to which I would now like to turn is just that: *What* is it about GIS and its use that makes it different both from work generally and from geography more specifically?

CONCEPTUAL ORDER IN GIS

In several recent works I have looked at contemporary phenomena—the rise of postmodernism (Curry, 1991c), the changing face of communication, and the concern with ethics in the discipline (Curry, 1991d)—and concluded that those phenomena are expressive of a kind of unmooring within the discipline, where critics have begun to see the traditionally related levels of research, communication, and social practice as coming apart. It seems to me that the development of GIS expresses the same unmooring, albeit in a different way. Hence, a starting point in understanding the ethical issues that face us in GIS is to look at each of these levels.

At the level of research, it seems to me, it is important to note three related conceptions in GIS, each of which would largely be accepted by its users. The first is that it is possible to locate events or objects in space, and that it is often important to do so. The second is that the basic aim of research is in part, perhaps in large part, the amassing of information, as opposed to the discovery of knowledge. The third is that there is something called rationality, involving the use of well-defined sets of rules, which can be adequately modeled. The important thing to note here, and I will come back to this, is that each of these three concepts—of space, of information, and of rationality—finds a specific expression in the context of the computer (see, e.g., McGranaghan, 1985; Goodchild, 1985).

At the level of communication, or writing and representation, a different series of assumptions is made. One assumption concerns the nature of language. It is typically assumed that one can map—here I am using the term metaphorically—language onto the world, just because there are linguistic particles that are analogous to the facts or objects of the world. This assumption in turn suggests that it should in principle be possible to incorporate local systems into ones that are increasingly universal. A related assumption is the notion of the "other." By this I mean that the language that is used to characterize the objects of one's research establishes a sharp distinction between those objects and the researcher. To put it more familiarly, when I talk about people within the context of a GIS I use different terms—and a different notion of agency—than I do when I speak with my friends and colleagues. Also, it is assumed that time—if time is considered at all—is a metric on a par with space; both can be treated as variables.

At the level of practice it should be obvious that more than elsewhere in geography, in GIS there is an attempt to "get out into the world." The earliest of these systems were designed not for pure research, but as instruments of policymaking, and this has remained true to today. Hence, GIS usually assumes a particular set of relations to society and to politics. These extend to notions of social change and political action, where the availability of good information is a means to good policy, and where technological and social change go hand in hand. As an aside, but an important one, I should note that in GIS, as elsewhere, thinking about knowledge and society is shot through with spatial metaphors. Not the least of these, and I have used it here just because it is so pervasive and compelling, is that there are several "levels," with research or data collection at the bottom, then writing and communication in the middle, and the social practice of science at the top.

In some respects these assumptions are quite familiar. Certainly, they predate GIS, and many can be found in the basic texts of the quantitative revolution in geography. As another aside, let me note that the persistence of these views, in the face of devastating criticisms in the history, philosophy, and sociology of science, as well as within geography, says something important about the nature of scientific change and the relationship between science and the social conditions within which it develops. That, though, is in part another issue. What *is* important is that these assumptions are in conflict, both within and between levels. The nature of these conflicts will now be my concern.

CONFLICT AND STRAIN: CONCEPTIONS OF MORALITY AND AGENCY IN GIS

Metaethics: Emotivism and Descriptivism

The first of these conflicts is at the metaethical level. I hasten to add that this strain is typical not just of works in GIS, but also in science more generally in the 20th century. When I say that it is at the metaethical level, I mean that it concerns the nature of statements within ethics, rather than their content. Most generally—and I am simplifying here—there are two possibilities; one may see ethical statements, like "This is good," as being very much like factual statements, or one may not. It seems to me that the strain here arises just because there is a tendency to hold both views at once.

In order to understand how this happens, it is helpful to look back to early in this century, when logical atomism and logical positivism, the forerunners of current versions of empiricism, were just coming into their own. At that time people like Bertrand Russell, A. J. Ayer (1946/1952), and C. K. Ogden and I. A. Richards (1923) began to develop a theory of meaning in language (see also Stevenson, 1948). According to that theory, which in some ways is grounded in the much earlier work of David Hume (1777/1975), statements that are meaningful must either be direct factual statements or they must be statements that are true by virtue of their meaning. If they are neither, and cannot be decomposed into such simple statements, linguistic utterances are just that, utterances, on a par with grunts and groans. For Ayer and the rest, ethical statements were just of this sort; they were expressions of emotion, and for that reason this metaethical view is termed "emotivism." Within the context of logical positivism this view seemed to make good sense. It heralded a day in which clear thoughts would be expressed clearly, and when the continental and idealist baggage, the sort of Victorian effluence, of philosophy would be superseded by a more modern view.

It seems clear that the attitude to language expressed in GIS leads directly to this view, since it sees language, at its best, as grounded in direct descriptions of the world, and hence—if only implicitly—devalues moral discourse (Hollis & Lukes, 1982). At the same time, when I say that I think that here there is a strain within geography and modern science, that is because we also find an implicit—and very different—metaethical theory. That theory sees it as absolutely true, in a *descriptive* sense, that science is good and that the current version of the scientific method is *the* scientific method. The nature and sources of this view are seldom laid out, but they typically involve denying that the value of science is simply a matter of opinion or even that it arises from the fact that science somehow "works." Whatever the sources of this view, it is in direct conflict with the emotivism espoused elsewhere, and the result is a strain within the discipline.

Teleology and Deontology in Moral Theory

A second strain is related to the nature of the moral theories that are being advocated. Here it is traditional to divide moral theories into two types. On the one hand, there are theories that measure the value of an act by its consequences; the most common of these teleological or consequentialist theories measure pleasure or happiness, and are termed utilitarian (Bentham, 1823/1948; Mill, 1863/1972). On the other hand are those theories termed deontological, which focus on obligations, and specifically on the moral agent as an autonomous being (Kant, 1785/1976). These theories see the consequences of an action as quite irrelevant to its moral worth. Quite to the contrary, they argue that an act is morally estimable only to the extent that its agent is acting out of a belief that the action is a good action, where good actions are defined as those that treat all humans as rational moral agents. Here, too, the strain within GIS is also found in geography, and in science more generally, although I think that it is fair to say that this kind of strain is not so general as the one discussed in the previous section. There is a strain because both views are held, and again they appear to be held simultaneously.

Consequentialist theories are abundant in geography. Cost–benefit analysis is, of course, one of the most obvious examples, and we find it in a wide range of subdisciplines. Much of the work done on hazards is based on this cost–benefit model. But so too are environmental impact analyses and distance minimization models. All measure the desirability of an action by comparing the good and bad consequences that may arise from it, and all gain their appeal from their appearance of neutrality and objectivity, and hence from the very values that science appears to support.

At the same time, both science and academics rely fundamentally on a notion of the individual as an autonomous agent. While scientists may well see themselves as operating within the context of a "system," they typically see their ability to be creative and to exercise their own scientific judgment as requiring that they be seen as autonomous agents. This view is implicit in every philosophy of science, which describes the methods that the scientist ought to use in distinguishing between truth and falsehood, and implicitly assumes that the scientist is independently able to exercise a form of agency in choosing. It is also implicit in most traditional histories of science, where science is seen as a progressive search for the truth, and as a process of shedding distorting and stultifying myths and traditions. Indeed, even the most adamant determinists argue that they have reached their beliefs as a result of freely choosing between alternative positions.

HETERONOMY AND AUTONOMY: THE EXPERT, THE INDIVIDUAL, AND THE OTHER

If there is a conflict between teleological and deontological theories, if scientists at once apply cost–benefit analysis while denying that it applies to them in some parts of their daily lives, there is another strain—one that will occupy me for the rest of this chapter. This is a strain *within* the deontological tradition, as scientists by turns see themselves as autonomous and deny that autonomy, either to themselves or more commonly to others. When one denies autonomy to oneself, when one claims that there is some other force in control of one's life, one is said to be heteronomous. Nowhere is this heteronomy more clear than in GIS, where the notions of rationality, of representation, of universalism, of space and time, and especially of technological development lead one to see the choices that one makes as *driven* in very particular ways.

In the case of technological development, and especially where computers are involved, this heteronomy is quite evident. We find it there in the way that one is always looking forward to the next step in computing power, with the assumption that there is an inexorable direction to this development, and that one has no real choice in the matter. This theme, of autonomous technology, or technology out of control, extends well beyond computers, to technology more generally—and it has not gone unnoticed (Winner, 1977). In the case of GIS this leads to a kind of fatalism, where discussions about the desirability of the development and use of the systems are seen as frivolous attempts to push back the tide.

In the case of GIS the importance of representation provides a similarly teleological force. This is because implicit in the notion of the visual representation of research results is a notion of accuracy as an unquestioned and unquestionable ideal, as an ideal that is therefore autonomous, its own driving force. The very idea of mapping seems, that is, to suggest that each point on the surface to be mapped ought to be unambiguously mappable. This is not to say that only in GIS do we find a desire for accuracy. It is, though, to say that in this case accuracy has become an end in itself.

Technology also can lead to heteronomous behavior in another way. As technological systems have become more complex, it has become increasingly difficult, and now is probably impossible, for anyone truly to understand the nature of any one system. This means that technological systems have become black boxes, and that our relations to them have an element of faith that they did not have before. This complexity has further consequences. First, it renders the whole notion of a teleological value system, and hence of cost–benefit analysis, problematic. With systems of this complexity, it is impossible to predict the results, and hence teleological value systems become impossible to apply (Wenz, 1983; Winner, 1977, 1986).

Second, the development of complex technological systems has tended to lead to an increased division of labor, within which different individuals understand parts of a system fairly well, but no single person understands the entire system. In the 20th century this division of labor has been manifested in two primary expressions of specialization and expertise (Hacker & Starnes, 1990): the Taylorism of the factory floor and the professionalization of certain sorts of experts, of which academics are one. I shall not dwell on Taylorism, but I would like to note in passing that recent movements, like the quality of work life movement on the factory floor and the similar white-collar movements sometimes associated with Japan and Japanese management, can in part be seen as attempts to escape the heteronomy inherent in Taylorism and its functionalist view of labor—and hence to escape the division of labor that we see operative in organizations that produce and use GIS. I do intend, however, to say a bit more about the matter of expertise.

The Development of Expertise

Most of us take it for granted that experts are needed; even most Luddites think that some people are better able to smash machines than others. There are, of course, exceptions. For example, Ivan Illich (1978) argues that the development of any forms of expertise or of the division of labor leads inevitably to the denial of personal autonomy, and hence must be avoided. As much as one might wish to object to utilitarian or teleological forms of moral theories, it does seem to me that here one is constrained to wonder whether the effects of such a change are at all what anyone *really* wants. So I think it better to assume that we do need experts, and then to ask what the consequences are for the issue at hand.

It is helpful here to distinguish between knowing *how* and knowing *that*. Knowing *how* refers to the ability to do something, the ability of the average person, say, to use a computer, to enter data, or to do analysis using simple, perhaps menu-driven, routines. Knowing *that* refers to knowledge about how something works. When we look at the role of technology in society, we find a somewhat complicated story. In the last few years some technological systems—and computers are a prime example—have obviously become much more complicated. Hence, for the average person, the ability to operate a system is increasingly a matter of knowing *how*, and decreasingly a matter of knowing *that*. It is less and less related to the understanding of how a system works, and technological systems look more and more like black boxes. A consequence of this, and an important one from the current point of view, is that this leads people to see themselves as using technologies that are, to use Langdon Winner's term, "autonomous"; to see those technologies as somehow driving change within society and providing the mechanisms, in a multiple sense, by which those changes are effected.

This leads, of course, to the need for experts who are able to understand the workings of those systems, both to design and to repair them. A complicating factor, though, is that the experts become expert at less and less. Indeed, it seems clear that for certain very large and complicated systems no single person can truly be said to "understand" the operations of the entire system technologically. Certainly that is even more the case when we extend our understanding of the technological system to include not just the hardware and the software, but also those who operate it, their families, and the associated economies and societies in which they live.

This implies—as I mentioned above—that cost–benefit analysis and other teleological moral theories are very nearly bankrupt; there is no way for an individual to really determine the results of a single action or set of actions, and to defer that decision to a team or group is merely to move the problem elsewhere.

There is another feature of the development of such systems, however, and that concerns the way in which—at least in the current context—the notion of expertise gets tied up with the notion of levels. I have already noted the pervasiveness of the use of the metaphor of "levels" in discourse about science, but that metaphor gets used more generally in the case of expertise, where it is related to the notion that society can be laid out in terms of different levels, with those who know more at higher levels and those who know less at lower levels.

The Notion of the "Other"

In earlier work on the development of the notion of expertise among students of the effects of nuclear war (Curry, 1986), I noted the way in which this has led to a "disenfranchisement" of, say, victims of the bombings of Hiroshima and Nagasaki, with an accompanying turn to the accounts of experts, and especially to those accounts couched in mathematical and system-analytic terms. In the case of GIS, we see a similar process in operation. Here the availability of massive amounts of information, especially in the context of the view that these systems are universalizable, leads to the notion that having this information provides one with a better understanding of the world.

Whether this is true is itself an open issue, but the belief that it is true has some unfortunate consequences. Most important, it leads experts to see those people to whom their data refer as "other." In part, the "other" is seen as existing in Cartesian space and technical, chronological time, rather than lived space, or place, and human or narrative time, while the person doing the judgment remains centered in the human world, seeing him- or herself as making decisions and acting freely. From the point of view of one who holds a deontological moral theory, this means that the "other" is fundamentally thought of as an object, rather than as a person, and hence is fundamentally treated unethically.

Furthermore, because the availability of information is seen as being of fundamental importance to the making of decisions, those who have that information see themselves as empirically better able to make decisions than are those who are merely "other." And this means that there are features of the use of these systems that are fundamentally antidemocratic.

Matters are, of course, complicated because those who run the systems are at the same time objects of inquiry, and in some respects are thereby forced to treat themselves as "others," to see their actions and behavior as something that occurs outside of themselves.

On Surveillance and Privacy

These accounts are antidemocratic in another way, too. This concerns the way in which the apprehension of the individual as an "other" relates to the desire to use GIS for surveillance (see Burnham, 1984; Laudon, 1977, 1986). We know, of course, that there has been a recent proliferation of national databases. The Internal Revenue Service has a database of taxpayers, and it has been used by other federal agencies to ferret out individuals who have failed to register for the draft, individuals who fail to repay student loans, and so on. The National Crime Information Center has a computerized criminal history system, and there are similar local systems. There are also a number of private systems, the most commonly known of which are credit reporting agencies. At the same time, companies engaged in mass marketing through direct mail have massive databases. The "low-end" companies have lists of all of the mailing addresses in the United States; when they can demonstrate that they have about 90% of those addresses the U.S. Postal Service provides them with the rest. "High-end" systems have names to attach to those addresses, and have been able through government agencies to compile reasonably extensive dossiers on individuals, on the cars that they drive, who lives in a household, and the like (Columbia Human Rights Law Review, 1973; Eaton, 1986; Freedman, 1987; Linowes, 1989; Weiss, 1988).

Here GIS provides an ability to go even farther; using voting records, home prices, and census data—to name only a few kinds of data—it is now possible to make educated guesses about any household's political and religious views, as well as its shopping preferences. Some companies—Fingerhut is one—have databases extensive enough that they send personalized catalogs; in principle, each household gets a catalog containing the items that its members are the most likely to buy.

I suppose that one can feel at least some relief that they have to send a catalog and cannot, instead, just go ahead and ship the items that they know that you do not have, will want, and can afford. But that seems like small recompense for a system that presumes to know so much about you. Indeed, what is perhaps most disturbing about this surveillance system—beyond the fact that it is largely unregulated—is that it presumes a notion of closure, a view wherein there is a population of individuals, and where it is possible to obtain measurable knowledge about each. It implies a truly closed society.

Individual and Community

One feature of such a society of others is that it treats the connections among individuals as merely contingent features, as artifacts of the system. Indeed, the presumption is that we live in a society made up in the first instance of individuals. I would argue that the very existence of such surveillance systems acts to accentuate previous trends in that direction. It does so at the same time that it leads to a disturbing reconceptualization of the nature of the individual. This is because the existence of a large-scale surveillance system is directly inimical to the existence of individual privacy, while it is privacy that gives people the opportunity to become individuals in the sense that we think of the term.

People define their individuality by selectively making public certain things about themselves. One important feature of this process involves memory; we rely on the fact that there are things about us that others can and will forget, and we are able, thereby, to feel that we live in a society where there is the possibility of redemption ("Privacy in the Computer Age," 1984; Schoeman, 1984). Yet a highly organized surveillance system denies that possibility and, in doing so, redefines who we are and who we can be. It is ironic, of course, that it also provides what has been termed a "license to forget"; people actually need to learn less and less, but they have access to more and more information (Winner, 1977). It is here that the relationship between knowledge and information appears most problematic.

This change in the conception of the individual extends, actually, to the practice of science. The sociology of science has been effective in showing the ways in which science relies on the development of customary forms of interaction, of community, and of trust. And these very concerns have been implicit in recent concerns within geography about a decline in ethical standards within the discipline. We find this in works by Brunn (1989) and Hanson (1988), journal editors who believe themselves to have witnessed a decline in honesty among authors, especially in the ways that authors publish similar, or even identical, pieces in several areas in their attempts to gain credit and prestige.

In part, actually, their concern is with the manipulation of computer databases, like the Social Science Citation Index; in a smaller and less harried discipline, one where people actually read one another's work before making judgments about its—and its author's—worth, the issue would not arise. It is of course slightly unfair to blame the technological systems for this tendency, and the unethical behavior that it allows; after all, there are all sorts of other reasons for the reliance by those in positions of authority on these indexes. And they do in some cases provide ready evidence of the visibility of individuals who are members of groups that are systematically discriminated against. Still, it remains true that the systems themselves establish an easy means for replicating the trappings of the traditional system of rewards, but in the absence of the traditional, communitarian bonds that might have allowed them to function.

If this change in the conception of the individual scientist or geographer is in part a result of the increase in the size of the discipline, it is also in part a result of the movement of geography into what Price (1963) termed "big science," science based on the use of rapidly changing technology, extremely expensive equipment, and large amounts of money. GIS are surely the first sign of this movement within geography. Such a development renders traditional modes of action problematic, especially when a discipline comes to be closely related with a demanding but unsophisticated public and a rapidly swinging door between academics and the private and government sectors. Within this context the geographer, too, becomes an "other."

CONCLUSION

I have argued that in various ways those who develop and maintain GIS adopt differing and even conflicting stances toward ethical issues. They adopt conflicting views of what an ethical statement is; they adopt conflicting views of the nature of ethical standards; and they apply to themselves standards that they do not apply to others.

Now, on the one hand, this is in itself an important conclusion. It suggests that to the extent that one hopes for guidance on policy issues from those who use these systems, one is likely to be disappointed because one finds oneself with a variety of conflicting answers. This is, of course, true across the sciences, and across portions of the human sciences. Yet here the case of GIS is special. While it is at least potentially possible that some approaches to the human sciences would avoid those contradictions, in the case of GIS this is not—for the reasons which I have enunciated above—possible. These contradictions are in a fundamental sense built into the systems.

One might, of course, wish to argue that these problems apply only to those who are directly involved in creating and maintaining the systems, and that those who simply use the output of the systems are somehow exempt from these criticisms. Unfortunately, this is not easy to do; to develop an understanding of the data adequate to a resolution of the problems which arise in the production of GIS would very likely render those systems irrelevant.

As I suggested at the beginning of this chapter, there is a second issue here, one raised by the very ethical inconsistencies of which I have spoken. Leaving aside the untoward social consequences of this inconsistency, one is likely, under certain views of the nature of the individual, to see this inconsistency as a product of a kind of decentering, even a dissolution of the "real" individual. It has been common, recently, to see this as a product of the postmodern era, where the individual comes to be no more than the sum of his or her roles. On this view, this process is to be criticized; it is contrasted with a situation wherein the individual is a "real person," a whole.

Yet as I have suggested in previous works, that view itself typically derives from a view of the individual as defined and driven by a kind of unitary mind, which orders—or ought to order—all of an individual's experience (Curry, 1989, 1991b). This view, from Descartes and Kant, raises a host of problems. I argued in those works—and what we have seen in the case of GIS supports this view—that it makes far more sense to see individuals as involved in a whole range of activities, each with its own set of grammars, logics, and systems of authority and justification. These patterns of activity are seldom explicitly defined, and they overlap in various ways. One way in which they overlap is that the same terms and phrases are often used in various ways, with inconsistent meanings. I would argue that in the case of GIS we see this very thing, as terms like "individual" and "good" appear in various contexts, appearing to link those contexts into a seamless web, when in fact there are fundamental differences.

The significance of these differences is twofold. First, they are, quite simply, differences, and they reflect the inconsistency to which I have pointed. Second, and more important, within each of these patterns of activity only certain kinds of consequences can be reached. In a discourse wherein one speaks of a person as an "other," one quite simply cannot treat that person as an autonomous individual. There are, that is, limits to what can be said (and done) within a given context.

It is typical to think of patterns of activity that are circumscribed by sets of discourses as lacking in solidity, as easily subject to change. But here the technological and representational systems that make up important elements of GIS rigidly define such contexts; for that reason the changes that may occur within those systems are more predictable than those that might occur within other contexts. Indeed, here sets of ideas have become so closely intertwined to the system of technology and representation that the discourse that is carried on within a given context can be seen as itself fundamentally limiting the sort of changes that may occur. I would argue that it is this which gives the inconsistency of ethical discourse in GIS its inevitability and its importance.

ACKNOWLEDGMENTS

Earliers versions of this chapter were delivered in departmental seminar series at San Diego State University and the University of Delaware. The author wishes to thank members of those departments for their comments. An earlier version of this chapter was published in the Departmental Monograph Series, Department of Geography, San Diego State University (Curry, 1991a). This research was funded in part by a grant from the Academic Senate of the University of California at Los Angeles.

NOTES

1. Anderson (1969), Brunn (1989), Curry (1991d), Hanson (1988), Kirby (1991), and Mitchell and Draper (1982) are the primary works. Brunn, Curry, and Hanson are quite specific; each deals with issues surrounding writing and publication.

2. In the matter of computers, see, for example, Forester and Morrison (1990), Hoffman and Moore (1982), Johnson (1985), Johnson and Snapper (1985), and Turn and Ware (1985). In science and the social sciences, see Callahan et al. (1983), Chambers (1980), Schmaus (1983), Trend (1980), and Weil and Snapper (1989). In professionalism, see Baumrin and Freedman (1983), Durkheim (1958), Goldman (1980), Ladd (1985), and Vernon (1987). In the case of technology the literature is quite vast; see especially Kranzberg (1980). In the case of business, see MacIntyre (1983); of publication, see Mahoney (1990) and the controversy surrounding his piece; and of representation, see Gross, Katz, and Ruby (1988), Harley (1990), and Warren (1980).

REFERENCES

Anderson, J. (1969). Moral problems of remote sensing technology. *Antipode, 1*(1), 54–57.

Ayer, A. J. (1952). *Language, truth, and logic* (2nd ed.). New York: Dover Books. (Original work published 1946)

Baumrin, B., & Freedman, B. (Eds.). (1983). *Moral responsibility and the professions.* New York: Haven Publications.

Bentham, J. (1948). *An introduction to the principles of morals and legislation.* New York: Hafner. (Original work published 1823)

Brunn, S. D. (1989). Editorial: Ethics in word and deed. *Annals of the Association of American Geographers, 79*(3), iii–iv.

Burnham, D. (1984). *The rise of the computer state.* New York: Vintage Books.

Burrough, P. A. (1986). *Principles of geographic information systems for land resource assessment.* New York: Oxford University Press.

Callahan, D., Jennings, B., Warwick, D. P., Pettigrew, T. F., Levine, C., & Bermel, J. (1983). Ethics and social inquiry. *Hastings Center Report, 13* (Special Suppl.), 1–16.

Chambers, E. (1980). Fieldwork and the law: New contexts for ethical decision making. *Social Problems, 27,* 330–341.

Columbia Human Rights Law Review. (Eds.). (1973). *Surveillance, dataveillance, and personal freedoms.* Fair Lawn, NJ: R. E. Burdick. (Original text published 1972, as pp. 1–235 of the *Columbia Human Rights Law Review, 4,*[1])

Curry, M. R. (1986). Beyond nuclear winter: On the limitations of science in political debate. *Antipode, 18*(3), 244–267.

Curry, M. R. (1989). Forms of life and geographical method. *Geographical Review, 79*(3), 280–296.

Curry, M. R. (1991a). *Morality and agency in geographical information systems.* (Departmental Monograph Series No. 91-1). San Diego, CA: San Diego State University.

Curry, M. R. (1991b). The architectonic impulse and the reconceptualization of the concrete in contemporary geography. In J. Duncan & T. J. Barnes (Eds.), *Writing geography: Text, metaphor, and discourse* (pp. 97–117). New York: Routledge.

Curry, M. R. (1991c). Postmodernism, language, and the strains of modernism. *Annals of the Association of American Geographers, 81*(2), 210–228.

Curry, M. R. (1991d). On the possibility of ethics in geography: Writing, citing, and the construction of intellectual property. *Progress in Human Geography, 15,* 125–147.

Durkheim, E. (1958). *Professional ethics and civic morals* (C. Brookfield, Trans.). Glencoe, IL: Free Press.

Eaton, J. W. (1986). *Card-carrying Americans: Privacy, security, and the national ID card debate.* Totowa, NJ: Rowman and Littlefield.

Forester, T., & Morrison, P. (1990). *Computer ethics: Cautionary tales and ethical dilemmas in computing.* Cambridge, MA: MIT Press.

Freedman, W. (1987). *The right of privacy in the computer age.* New York: Quorum Books.

Goldman, A. H. (1980). *The moral foundations of professional ethics.* Totowa, NJ: Rowman and Littlefield.

Goodchild, M. F. (1985). Questions, tools or paradigms: Scientific geography in the 1980's. *Ontario Geographer, 25,* 3–14.

Gross, L., Katz, J. S., & Ruby, J. (Eds.). (1988). *Image ethics: The moral rights of subjects in photographs, film, and television.* New York: Oxford University Press.

Hacker, S. L., & Starnes, C. E. (1990). Computers in the workplace: Stratification and labor process among engineers and technicians. In S. Hacker, *"Doing it the hard way": Investigations in gender and technology,* D. E. Smith & S. M. Turner (Eds.), (pp. 175–194). Boston: Unwin Hyman.

Hanson, S. (1988). Soaring. *Professional Geographer, 40,* 4–7.

Harley, J. B. (1990). Cartography, ethics, and social theory. *Cartographica, 27*(2), 1–23.

Hoffman, W. M., & Moore, J. M. (1982). Ethics and management of computer technology. *Proceedings of the Fourth National Conference on Business Ethics.* Cambridge, UK: Oelgeschlager, Gunn, and Hain.

Hollis, M., & Lukes, S. (Eds.). (1982). *Rationality and relativism.* Cambridge, MA: MIT Press.

Hume, D. (1975). *Enquiries concerning human understanding and concerning the principles of morals* (3rd ed.). Oxford: Oxford University Press. (Original work published 1777)

Illich, I. (1978). *The right to useful employment and its professional enemies.* London: Lothian Pty.

Johnson, D. G. (1985). *Computer ethics.* Englewood Cliffs, NJ: Prentice-Hall.

Johnson, D. G., & Snapper, J. W. (Eds.). (1985). *Ethical issues in the use of computers.* Belmont, CA: Wadsworth.

Kant, I. (1976). *Foundations of the metaphysics of morals.* (L. W. Beck, Trans.). Indianapolis, IN: Bobbs Merrill. (Original work published 1785)

Kirby, A. (1991). On ethics and power in higher education. *Journal of Geography in Higher Education, 15,* 75–77.

Kranzberg, M. (Ed.). (1980). *Ethics in an age of pervasive technology.* Boulder, CO: Westview Press.

Ladd, J. (1985). The quest for a code of professional ethics: An intellectual and moral confusion. In D. G. Johnson & J. W. Snapper (Eds.), *Ethical issues in the use of computers* (pp. 8–13). Belmont, CA: Wadsworth.

Laudon, K. C. (1977). *Communications technology and democratic participation.* New York: Praeger.

Laudon, K. C. (1986). *Dossier society: Value choices in the design of national information systems.* New York: Columbia University Press.

Linowes, D. F. (1989). *Privacy in America: Is your life in the public eye?* Urbana: University of Illinois Press.

MacIntyre, A. C. (1983). Why are the problems of business ethics insoluble? In B. Baumrin & B. Freedman (Eds.), *Moral responsibility and the professions* (pp. 350–359). New York: Haven Publications.

Mahoney, M. J. (1990). Bias, controversy, and abuse in the study of the scientific publication system. *Science, Technology, and Human Values, 15*(1), 50–55.

McGranaghan, M. (1985). Pattern, process, and a geographic language. *Ontario Geographer, 25,* 15–27.

Merton, R. K. (1973). *The sociology of science: Theoretical and empirical investigations.* Chicago: University of Chicago Press.

Mill, J. S. (1972). *Utilitarianism.* London: Dent. (Original work published 1863)

Mitchell, B., & Draper, D. (1982). *Relevance and ethics in geography.* London: Longman.

National Center for Geographic Information and Analysis (1989). *Introduction to GIS: NCGIA core curriculum.* Santa Barbara: National Center for Geographic Information and Analysis.

Ogden, C. K., & Richards, I. A. (1923). *The meaning of meaning.* London: Kegan Paul.

Price, D. J. de S. (1963). *Little science, big science.* New York: Columbia University Press.

Privacy in the computer age. (1984). *Report from the Center for Philosophy and Public Policy, 4*(3), 1–5.

Schmaus, W. (1983). Fraud and the norms of science. *Science, Technology, and Human Values, 8,* 12–22.

Schoeman, F. (Ed.). (1984). *Philosophical dimensions of privacy: An anthology.* Cambridge: Cambridge University Press.

Star, J., & Estes, J. (1990). *Geographic information systems: An introduction.* Englewood Cliffs, NJ: Prentice-Hall.

Stevenson, C. L. (1948). The nature of ethical disagreement. *Sigma, 2,* 469–476.

Trend, M. G. (1980). Applied social science research and the government: Notes on the limits of confidentiality. *Social Problems, 27*(3), 342–349.

Turn, R., & Ware, W. H. (1985). Privacy and security issues in information systems. In D. G. Johnson & J. W. Snapper (Eds.), *Ethical issues in the use of computers* (pp. 133–147). Belmont, CA: Wadsworth.

Vernon, N. D. (1987). Toward defining the profession: The development of the code of ethics and the standards of professional practice of the American Society of Landscape Architects, 1899–1927. *Landscape Journal, 6*(1), 13–20.

Warren, C. A. B. (1980). Data presentation and the audience: Responses, ethics, and effects. *Urban Life, 9,* 282–308.

Weil, V., & Snapper, J. W. (Eds.). (1989). *Owning scientific and technical information: Value and ethical issues.* New Brunswick, NJ: Rutgers University Press.

Weiss, M. J. (1988). *The clustering of America.* New York: Harper and Row.

Wenz, P. S. (1983). Ethics, energy policy, and future generations. *Environmental Ethics, 5,* 195–209.

Winner, L. (1977). *Autonomous technology: Technics-out-of-control as a theme in political thought.* Cambridge, MA: MIT Press.

Winner, L. (1986). *The whale and the reactor: A search for limits in an age of high technology.* Chicago: University of Chicago Press.

Zuckerman, H. (1984). Norms and deviant behavior in science. *Science, Technology, and Human Values, 9,* 7–13.

CHAPTER 5

Computer Innovation and Adoption in Geography

A CRITIQUE OF CONVENTIONAL TECHNOLOGICAL MODELS

Howard Veregin

TECHNOLOGICAL NEUTRALITY AND TECHNOLOGICAL VALUE

The computer revolution has inspired a great deal of interest in the social impacts of computer technology and the ways in which it has transformed and reshaped human experience with the world. These changes have been described and analyzed in numerous scientific and popular works presenting competing visions of the technological world of the future. Advocates see the computer as a means of achieving a more democratic society and promoting a broader base of understanding of current social, political, and economic issues. These effects are attributed to the ability of the computer to facilitate access to information without the bureaucratic and institutional costs associated with more traditional information sources. Detractors view such claims with skepticism, arguing that information is not the same as knowledge, and that the advent of the computer has led to other, less desirable, changes that have too frequently been overlooked. The focus of this debate has shifted repeatedly as the technology has evolved and developed, and as various scientific and popular fashions have come and gone.

The computer has become a ubiquitous feature of many of the institutions with which we interact in our daily lives. However, because this technology is of relatively recent origin, many of its more significant impacts are only now beginning to emerge and be recognized. This is no less true in academic disciplines such as geography, where the computer is being adopted and utilized for an ever broader range of research and educational purposes. Even the most casual perusal of the literature reveals that the computer has become an integral component of much geographical research. For example, the computer revolution has fueled much of the growth in quantitative geography. Studies show that the use of quantitative methods in articles published in geographical journals has increased by approximately 50% over the last 30 years (Slocum, 1990). Quantitative geography is now considered "mainstream" and many of the discipline's most influential journals are predominantly quantitative in nature.

Of course, quantitative geography is not the only area in which computer applications are important. Many journals and annual conferences now cater almost exclusively to research pertaining to computer applications in cartography, remote sensing, and spatial data analysis. Membership in technical specialty groups in some professional geographical societies has burgeoned in recent decades, often at the expense of systematic and regional areas of expertise (Goodchild & Janelle, 1988). Nowhere is the growing importance of the computer in geography reflected more strongly than in the field of geographic information systems (GIS), which has over the past decade witnessed an enormous increase in levels of professional interest, funding, research, education, and training (National Center for Geographic Information and Analysis, 1989). Adoption of the computer has also enhanced geography's linkages to the public and private sectors, where geographers familiar with computational techniques are assuming a practical and profitable role. Classroom instruction in computational techniques is likewise becoming increasingly common in geography departments, and the mastery of a variety of technical skills—from statistical methods to remote sensing, GIS, and computer-assisted cartography—is fast becoming an essential component of graduate and undergraduate degrees in geography.

Few would deny that the computer revolution has had a profound impact on the character of geography as an academic discipline. Moreover, because the rate of technological innovation and adoption shows little sign of receding, at least in the immediate future, the forces of change thus set in motion will continue to shape the discipline for some time to come. Judging from published literature on the topic, however, most geographers apparently conceive of the computer as little more than a tool, a technological apparatus that allows them to perform the same tasks they performed in the precomputer era, but now with more speed and greater flexibility. Those who embrace the computer speak of the great improvements in efficiency that it brings to data handling and processing (e.g., Kellerman, 1983; Moellering & Stetzer, 1983; Monmonier, 1983; Maguire, 1989). The computer places more facts in the hands of the geographer, permits data analysis to be performed more easily and quickly, and allows the geographer to "conduct familiar operations much faster and for much larger areas to test existing theory and to generate new hypotheses" (Dobson, 1983, p. 138).

From the vantage point of this tool-based model, the computer is merely an extension of existing human capabilities. Accordingly, the computer does not portend any significant qualitative changes in thought or action, and its impacts are limited to issues of speed, efficiency, and processing power. Tools are typically construed as intrinsically neither good nor evil, that is, they are viewed as neutral or value-free. Technological value is instead furnished by the use and application of a tool, as rooted in a specific cultural context. Thus a handgun may be used for socially constructive purposes (protection and defense) or socially destructive purposes (violent crime), but the handgun itself possesses no values. Likewise, the computer is viewed as value-free, such that its net impact depends only on the manner in which it is applied. Computers have thus been described as "essentially neutral in regard to many of the terms that geographers have used to characterize their work" (Dobson, 1983, p. 136), and the use of computers in geographical research has been characterized by many as inherently no more or less humanistic than its manual counterpart.

However, the notion that technology is value-free is inherently misleading. While not actually possessing values as inherent properties, technologies nevertheless prescribe certain ranges of action, use, or purpose that in effect define technological value. In one sense, for example, the private automobile is simply a technological apparatus that reduces travel time between distant places. At the same time, the driver of the automobile is prevented from traveling along routes that might be used when walking or cycling, is forced to select among a finite set of appropriate paths of travel, is prevented by law from moving too quickly or too slowly, and is closed off from immediate contact with the environment. The automobile owner is also forced to accept the responsibility of ownership, which implies the need to utilize a host of technological contrivances for fueling, maintenance, and service. Moreover, in some parts of the world, the near-universal acceptance of the automobile has led to a dispersed pattern of settlement in urban areas and a profusion of freeways connecting isolated suburban enclaves to the workplace. Indeed, the automobile has helped to reshape the morphology of urban areas in such fundamental ways that it is difficult to imagine rebuilding these areas in a form that is less dependent on the availability of private means of transportation.

Paradoxically, the most significant impacts of technology tend to occur when the technology becomes indistinguishable from the fabric of everyday life (Weiser, 1991). This disappearance reflects a psychological adaptation process in which it is no longer necessary to be conscious of actually using the technology in order to achieve some goal. Thus the most profound technologies are those that are so ubiquitous and commonplace that they are unconsciously accepted as somehow indispensable. The private automobile again provides a good example. If the automobile was beyond the economic means of all but a handful of individuals, or if the technical complexities were such that the automobile proved to be impossible to maintain and repair, then the effects of this technology as sketched above would clearly not have occurred. Indeed, low cost and ease of maintenance have contributed directly to the ubiquitous character of the automobile and, at least in North America, there seems to be scarcely an individual who does not consider it to be an essential component of life. Meanwhile, the changes wrought by the automobile reinforce this belief, as it becomes increasingly difficult to work, shop, and entertain oneself in an environment built under the assumption of the automobile's universality.

Technology's tendency to disappear into the background of everyday experience is essential to its ability to transform human thought and action, and to reshape human experience with the world. Technology is all around us, although we often do not perceive it as such. Every human contrivance is by definition technology. However, the popular conception of technology as "high-tech" electronic and mechanical gadgetry ignores the technology of earlier eras, much of which is still in use. How often, for example, is a simple dinner fork described as technology? And yet this seemingly innocuous device has transformed not only what we eat and how, but also the details of mealtime interaction and, by extension, whole patterns of social etiquette and protocol. The culture of good manners spills over into a larger milieu of social graces and accepted customs that were clearly unrealizable in the era before the fork was invented, and in cultures where the fork is not used as a contrivance to facilitate dining. Similar conclusions may be reached with reference to other technologies as well.

The most significant impacts of technology therefore tend to occur when the technology has ceased to be a "tool" and has become an integral and indispensable component of everyday life. It is precisely for this reason that the tool-based model of technology is inadequate as the basis for judgments about technological impact assessment. In terms of the rationale for adoption of computer technology in geography, the model offers little by way of enlightenment. Apart from a simple enhancement of data handling and processing capabilities, the model proffers no other benefits of the computer. The tool-based model has even less to say about the negative impacts of this technology. With some exceptions (e.g., Pickles, 1991), these negative impacts have not been studied systematically in the geographical literature. Indeed, there seems to be an unwillingness on the part of geographers to admit that significant undesirable effects may be associated with the adoption of computer technology.

Moreover, the negative impacts that have received the most attention in the geographical literature tend to be fairly innocuous problems, such as "mindless number crunching" (Dawson & Unwin, 1977). A frequent criticism is that the advent of the computer leads to the development of a "computer culture" whose language and practices many find inaccessible and objectionable. Those who use computers in geographical research are also sometimes characterized as mere "button pushers" who lack a real understanding of the complexities of geographical phenomena. The implication of this form of criticism is that computer applications are technological games and diversions rather than components of "true" geographical research. This viewpoint is perhaps reinforced by the tendency for geography departments to label courses in computer applications as "technical"—the implicit assumption being that these courses train students in skills that they can bring to bear in other, substantive areas of geographical analysis.

What this assumption ignores is that, in these technical areas, theory and practice are inseparable. The computer is both a technology and a means for describing geographical phenomena. Computer applications imply a change in the relationship between the geographer and the worlds that he or she wishes to describe and explain. The computer enables new aspects of these worlds to be seen, while simultaneously blocking other vistas from view. The result is the development of a new language for describing geographical phenomena that may bear little relationship to the languages adopted in substantive areas of study that technique is assumed to service. Those who would argue that computer users are mere technicians rather than "real" geographers should not forget that all technologies, from pencil to printed book to statistical methods, simultaneously constrain and liberate the range of thought and action that one may bring to bear in their application and use.

THE COMPUTER AS A FORCE FOR INNOVATION

Before focusing attention on some of the negative impacts of the computer that have not been addressed in the geographical literature, I would like to reexamine some of the positive impacts of this technology in light of the limitations of the tool-based model. According to this model, the main benefits of the computer derive from the enhancements in efficiency it brings to data handling and processing. There are, of course, countless examples of techniques practiced in the precomputer era that have benefited enormously from the introduction of computer technology. Many of these techniques are labor-intensive, as in the case of manual cartography and statistical analysis. Although statistical analysis was carried out by geographers long before the widespread adoption of the computer, manual methods of analysis were slow, cumbersome, and fraught with error. Time and cost requirements combined to prohibit multivariate analysis or the use of large databases, at least to the degree that these are utilized today. Simple "number crunching," being a straightforward mapping of one real number space onto another, also lends itself rather easily to sequential sets of programming instructions. It is not surprising, then, that statistical analysis was one of the first analytical methods to which computing systems were applied in geography, or that it continues to be one of the computer's primary application areas.

However, there are other, more fundamental, impacts that must also be addressed. These impacts derive from the ability of the computer to serve as a catalyst for innovations in the understanding of geographical phenomena. On one level this occurs when the capabilities of computer systems are extended beyond emulation of existing practices. Here the computer suggests new approaches for selecting, recording, transforming, and integrating data. On a second level, through a transformation of the relationship between data and model, the computer expands the range of geographical phenomena that can be analyzed and studied. In simulation modeling, for example, data are an integral component of model formulation, rather than being used only for the purposes of validation and calibration. This has important implications for the modeling of systems whose behavior is inherently complex, uncertain, or chaotic.

As I noted above, many of the more significant impacts of technology derive from its tendency to recede into the background of everyday experience. When this occurs, one is freed to use the technology without it being an encumbrance, and one is able to focus on the goals that the technology can be used to achieve. Consider the case of computer-assisted cartography. Many of the automated techniques in this field were initially designed to emulate what had been done previously in a manual environment. The possibility of eliminating many of the repetitive and labor-intensive tasks of manual cartography using a digital approach was achieved by the early 1980s (Goodchild, 1988). Subsequent insights have leaped beyond the emulation of manual practices as computer technology has become more commonplace in cartographic production, research, and education. Current paradigmatic approaches in cartography, including exploratory data analysis, multiple representations of thematic variables, and dynamic and linked displays (MacEachren & Monmonier, 1992), are possible only because cartography has been liberated from the time and cost constraints associated with manual techniques. In a computer environment one can freely manipulate and integrate data, generate visual representations, and discard these at will. Other innovative cartographic display capabilities—including three-dimensional representations of spatial distributions, cartograms, trend surfaces, and maps of time-, cost-, and cognitive-space—are all made possible, to a greater or a lesser degree, by the extension of computer technology beyond traditional manual approaches.

More important, perhaps, is the development of alternate models of geographical phenomena due to the computer's power to act as a metaphor for the real world. An example is the raster data model, in which the spatial distribution of some phenomenon is represented by assigning values to an ordered set of nonoverlapping cells. This model mirrors the way in which information is displayed on the computer screen. As Goodchild (1988) notes, the raster model represents a radical departure from the vector model associated with traditional cartography. Indeed, if traditional cartography had evolved in a world in which the raster model was already firmly established, many of the concepts we now hold to be self-evident, such as the internal homogeneity of mapping units, might never have taken hold. That is, our interpretation of geographical phenomena is in part a function of the cartographic language that may be brought to bear in describing these phenomena.

Computer-based simulation models of geographical phenomena stand in stark contrast to models developed in the precomputer era. The preponderance of deductive logic in much traditional geographical modeling (e.g., Abler, Adams, & Gould, 1971) derives from a conception of the scientific method in which empirical observation is performed selectively based on existing theory (Openshaw, 1989). Geographers have until quite recently lacked a consistent and convenient mechanism for developing inductive models on a large scale, owing to the absence of computing systems for spatial data manipulation and analysis. In simulation modeling, data and model are inseparable, because the model of the system under consideration is a complete, step-by-step description of the evolution of that system over time. In this context, inductive logic becomes a primary mode of scientific inquiry.

This reorientation of the value of data also has important implications for the modeling of systems whose behavior is inherently complex, uncertain, or chaotic. These systems are unique in the sense that they can evolve into radically different states as a function of relatively minor variations in initial configurations and transition rules. Simulation modeling represents a formal mechanism for developing an understanding of such systems that was not possible in the precomputer era. In contrast to many other spatial analytic techniques that focus on spatial pattern, simulation modeling is inherently process oriented and recognizes that the same spatial pattern may be produced by different processes, and that the same process can give rise to qualitatively different patterns. There are numerous examples of spatial processes for which simulation modeling might usefully be employed. Perhaps the most important are those global environmental problems, such as stratospheric ozone depletion and greenhouse warming, that require large-scale simulation modeling and a rich supply of data covering a range of environmental parameters on a global scale.

REVERSE ADAPTATION AND THE COMPUTER DATABASE

Paradoxically, the primary negative impacts of the computer derive from the same source as its main positive impacts. As a language for describing geographical phenomena, the computer opens up new vistas while simultaneously closing off others from view. Like all technologies, the computer prescribes certain ranges of action, use, and purpose as a function of its characteristic properties. These prescriptive tendencies occur in a variety of contexts. On the most basic level, the characteristics of computer systems prescribe the manner in which humans relate to them. Human interaction is required to ensure the continued functioning of these systems, and to use and apply them in problem-solving activities. In the 40-odd years that the computer has been available for use by business and academic institutions, patterns of human interaction have changed as a function of the evolution of computer technology itself. The large mainframe computers that prevailed during the 1950s and 1960s required bureaucratized maintenance and control structures, including a centralized location, a large amount of physical infrastructure, and an army of skilled technicians for system monitoring, operation, and maintenance. In both real and metaphorical senses, users were isolated from these systems, and were able to interact with them only in limited ways according to prescribed and codified communications mechanisms.

In the 1970s and 1980s the nature of this interaction changed through the development of mini- and personal computers. Personal computers in particular require little in the way of physical infrastructure, and are simple enough to be operated and maintained by any educated user. The characteristics of these systems thus permit direct control by the user, without the filter of bureaucratized control structures. Interestingly, in the 1990s these trends are again reversing. Workstations deliver the power of a mainframe to the desktop, but recent trends in networking imply a resurgence in the isolation of control in the hands of the minority possessing expert knowledge, due to the complexities associated with the operation of distributed computing systems.

Even in the case of personal computers, of course, only limited ranges of action and use are permitted. The computer demands instructions of a certain type and in a certain order. The data on which these instructions operate—whether a file of text or a map of land cover types—must be organized and structured according to a logic imposed not by the demands of the problem to be solved, but by the nature of the computer's storage media and the functioning of its logical circuits. Inasmuch as "data" represent an abstract model of some real phenomenon, the structure and order imposed on data by the computer in effect define a language for describing and interpreting the world. This language may be enormously liberating, but like any language it simultaneously constrains the user by defining the boundaries of potential thought and action.

The prescriptive tendencies of the computer create a dynamic of "reverse adaptation" (Winner, 1977a). Reverse adaptation refers to the transformation of existing goals to accommodate a new technical means. Goals are in effect rearranged in accordance with the demands of the technological order. In extreme cases, the broader social context ceases to be relevant as long as technological demands are satisfied and maintained. Reverse adaptation adds a twist to the conventional view that the computer allows the geographer to perform familiar operations more quickly and efficiently. It also contradicts the prevailing notion of the computer as inherently neutral (Cromley, 1983). Technology itself plays a crucial role in prescribing the avenues of research that may appropriately and feasibly be explored. The computer influences the way in which research problems are selected for study and the character of the assumptions, languages, techniques, and models brought to bear on the problem. This transformation amounts to a reorientation of research such that it is more in line with the new technical apparatus. The study of problems not amenable to automation may be abandoned, the study of others may be restructured for efficient data handling, and new problems may be adopted that focus explicitly on technical issues.

Reverse adaptation appears in a variety of contexts. It may occur during the collection and encoding of data. The function of a conceptual data model is to impose order and structure on the multiplicity of relationships between entities in the real world, in order to identify and extract relationships of interest. Due to the computer's enhanced capabilities for manipulating digital data, this process is usually described as "digitizing." Despite its innocuous name, however, digitizing is more than just the simple act of translating information into a form that is readable by a computer. Indeed, the digitizing process can exert a considerable effect on the nature and form of the computer database. In effect the digitizing process acts as a filter or constraint on the development of a model of some relationships presumed to exist in the real world, and as such is part of a language used for representing the abstract version of the real world in a computer database.

A nongeographical example may serve to clarify this point. Computer analysis of questionnaire data has, on the one hand, enormously expanded our abilities to assess public opinion on a variety of issues. On the other hand, questionnaires tend to abstract real-world opinion in terms of numerical indices amenable to computer encoding and analysis using spreadsheets and statistical software. Such indices may impose on public opinion an inordinate degree of rigidity and formality (Winner, 1977a). For example, the category labeled "other" is often reserved for responses to statements of belief that do not fall unequivocally into the two acceptable categories of "agree" and "disagree." Clearly, members of the "other" category actually exhibit wide differences of opinion. Indeed, the variation within this one category may reveal more about the range of opinion about the issue in question than simple tabulations of the percentages of agreement and disagreement.

Similar problems also apply in the context of spatial data. Waters (1989) provides the example of a computer database consisting of all known earthquake locations, except those earthquakes whose geographical coordinates were uncertain (as a function of their age); the latter, because they were not amenable to digitization, were simply ignored. Here potentially useful information was discarded due to the difficulty of digitizing an imprecise point location. How common have such practices become? Computer databases, like any model of geographical reality, are based on abstraction and generalization. Unfortunately, the exuberance with which the geographical community has embraced the computer has not fostered a critical awareness of the ways in which the computer prescribes how abstraction is to be performed. Rather, data are bent, shaped, and tweaked to fit them into a predefined mold. Similar concerns have been raised by critics of the quantitative revolution, who point out that statistical methods are based on restrictive assumptions about data that often cannot be satisfied in the context of spatial data.

In a GIS database, geographical features, or real-world entities, are defined as spatial objects. These objects acquire meaning only after they have been associated with aspatial or thematic information. Through this association, a GIS database achieves a representation of the multiplicity of relations among real-world entities. However, the thematic attributes associated with features in the database represent only a small subset of the attributes associated with the corresponding real-world entities. Thus a GIS database is an abstraction of the real world, necessarily incomplete and generalized. The manner in which this process of abstraction is carried out reflects a particular set of ontological concepts related to the manner in which knowledge about the world is formalized. The properties that supposedly best represent a reality unknowable in its totality are rooted in a specific cultural context and scientific paradigm.

These properties are, however, also filtered and modified as a function of technological feasibility and necessity. For example, this data model necessarily assumes that the features contained in a database represent unambiguous real-world entities. This assumption owes much to the evolutionary linkages between GIS and cartography. The three basic cartographic feature types—point, line, and area—are viewed as representative of real-world entities even though they are abstractions of a world more complex than this simple categorization would allow. Such abstractions are appropriate in cartography, where the emphasis is on effective communication of complexity through generalization, but in GIS analysis it is necessary to confront the errors that this data model induces through its inability to represent the inexactitude and uncertainty associated with most geographical phenomena.

Indeed, the notion that the features in the database represent unambiguous real-world entities is appropriate only for limited types of geographical data, such as the locations of census tracts, national boundaries, and other legal, administrative, or mathematically defined entities. Many other types of data, including soil, vegetation, and land-cover classes, exhibit neither recognizable areal boundaries nor strong spatial homogeneity within a given areal unit. Rather, these types of data are characterized by spatial heterogeneity and gradual transitions from one zone to an adjacent one. The cartographic representation of these phenomena—the so-called area class map—is based on generalization methods appropriate to the cartographic objective of communication of complex information. In the area class map, spatial variation associated with the phenomenon under consideration is minimized through a mapping of this variation to a generalized set of nominal classes. Operationalization of the area class map in a GIS produces a database exhibiting strong scale dependence. Mapping units may be attributed with an index of homogeneity, often referred to as "uniformity" or "purity," while artificial concepts such as the minimum mapping unit are invoked to define a minimum areal unit size that is recognizable as unique at a certain spatial scale.

Technological constraints affect the nature of the models of geographical phenomena that may be developed and applied in a technological environment. Often, the goals ostensibly guiding the application of technology are subsumed within the technological apparatus itself. This can give rise to actual decline in capabilities, as noted by Blakemore (1985) in terms of the quality of published maps, and by Batty (1988) in terms of the range of functionality in GIS systems for planning. Unfortunately, the continued evolution of computer systems does not suggest a lessening of this effect, since the problem arises from technological adoption itself and the concomitant restructuring of the ends to accommodate the technological means. Nor does the solution to the problem lie in searching for better ways to incorporate nontechnical goals within the domain of automated procedures (e.g., Dobson, 1983), as this in fact implies that these goals must be adapted to the technological means.

PASSIVE DISENGAGEMENT

Reverse adaptation—the transformation of existing goals to accommodate the new technical means—is also manifested in terms of disengagement between subject and object. The advent of new technical means may result in an increase in psychological distance between the individual and the environment in which the individual's actions are carried out (Borgman, 1982). An example is the psychological numbing that occurs on the modern electronic battlefield, where responsibility for decisions no longer rests with a human actor, but with the electronic apparatus mediating between the actor and the real world (Weizenbaum, 1976). Winner (1986) likewise maintains that public participation in the political process has steadily declined as the television has replaced face-to-face politics as the dominant mode of interaction and information dissemination. Passive monitoring of news and information via electronic media allows citizens to feel involved while simultaneously suppressing their desire to play an active role.

In geography, these tendencies may be interpreted in terms of a change in the patterns of interaction between the geographer and the worlds that he or she wishes to describe and explain. The computer promotes a remote, detached view of the world as seen through the filter of the computer database. Intimate knowledge of the world recedes into the background of "ground truth" as the computer screen becomes the medium through which the geographer interacts with the world. Porteous (1986) argues that the abandonment of foreign area research by geographers in favor of a technologically based, remote-sensing approach has led to a marked decline in knowledge of, familiarity with, and understanding of the world. Porteous lays the blame squarely on the shoulders of the remote sensing paradigm, which he characterizes as clean, cold, and detached. The necessary complement, which he refers to as "intimate sensing," is difficult, dirty, complex, and often dangerous, but at the same time rich, warm, and involved. Others have likewise warned against the pursuit of "macrolevel" generalization without the integration of "microlevel" knowledge (e.g., Rundstrom & Kenzer, 1989). Geographers need direct knowledge (knowledge of) as well as indirect knowledge (knowledge about).

This problem may also surface in the context of spatial decision support systems (SDSS), the application of specialized GIS technology in the decision-making arena. Advocates see SDSS as a means of providing timely and cost-effective information for policy decisions. SDSS also facilitates examination of a variety of possible scenarios in which the potential impacts, both positive and negative, of a particular policy decision can be assessed. However, little—if any—attention seems to have been directed at the question of how the technological apparatus disengages decision makers from the implications of adopted policies. For example, are decision makers less likely to incorporate subjective values and humanistic concerns in policy decisions because of the seeming remoteness of the world in which these policies have an impact?

This problem is reinforced by the fact that the data used in decision support are often derived from a variety of public and private agencies with a diverse range of mandates. As a result, there is little standardization or consistency in the means of data acquisition, classification, and representation. It has been suggested that the separation of data collection and data analysis has led to a lack of concern for data quality issues and the substance of geographical inquiry (Poiker, 1983; Rundstrom & Kenzer, 1989). In effect, the remote paradigm places a filter between the worlds that geographers wish to explore and the descriptive and explanatory models that they construct. This fact is not lost to other disciplines, such as history, where primary data sources are seen as essential in arriving at valid interpretations of historical events. Since historians also utilize secondary sources, it would appear that, at least to historians, the intimate and remote paradigms are complementary because each provides insights denied by the other.

GIS AS A UNIFYING FORCE

The notion that the computer is a tool that allows existing operations to be carried out more efficiently leads quite naturally to the conclusion that the computer's role in geography should be enhanced still further. As a valuable tool servicing the pressing needs of research, education, and practical training for employment in the "real world," the computer is viewed as the key to maintaining a competitive advantage over other disciplines in the war over funding and resource allocation. To some, the technology is critical to the very survival of geography itself. Geography's future as an academic discipline is seen to be closely linked to the ability of geographers to develop practical, real-world applications of geographical models that are in touch with contemporary issues and needs. As Openshaw (1989) notes, the computer facilitates the development of such models by providing a data-rich environment conducive to model testing and prediction.

At the same time, the computer is often seen as a means of establishing commonalities between geographers in a heterogeneous assortment of subfields. The diverse character of modern geography, it is argued, means that geographers are increasingly unable to communicate with each other and with those outside the discipline. According to this view, geography suffers from a "divisive proliferation of subparadigms" (Lundberg, 1989, p. 272). The discipline appears to be an ad hoc collection of economists, sociologists, planners, physical scientists, ecologists, statisticians, and, increasingly, engineers and computer scientists. In the face of such diversity, it is difficult to identify geography's primary purpose and explain to those outside the discipline how geography is unique and why it is even necessary. Modern geography, it has been argued, exhibits "overspecialization in areas that are more properly regarded as subfields of other disciplines" (National Center for Geographic Information and Analysis, 1989, p. 14), and geographers active in these subfields cannot hope to compete against the disciplines into whose domains they are intruding. According to this view, the distinct character of geography has been lost through the pursuit of too many narrow fields of inquiry, and the demise of geography as an academic discipline is a likely consequence. Therefore, as geographers, we ought to "reduce our emphasis on our too-numerous specialties and diverse interests" and instead "refocus our attention on the core of our discipline" (Abler, 1987, p. 516).

To many, computer technology provides the means to achieve this goal. The key technology here is GIS, since it is designed to handle spatial data, the common ground upon which all geographers presumably walk. GIS, so the argument goes, provides "a foundation for geographical methodology" (Tomlinson, 1989, p. 248)—an environment in which geographical research can be conducted based on a consistent and standard set of practices. GIS allows geographers to integrate diverse types of data over different spatial scales from the regional to the global, while the advanced capabilities of GIS for organizing and displaying these data transform the geographer's view of the world (Tomlinson, 1989). GIS, it is argued, provides the basis for investigating many of the complex, multivariate systems that have yet to be successfully modeled by geographers (National Center for Geographic Information and Analysis, 1989). It fosters communication and cooperation by providing an environment in which all geographers share the same concerns and cope with the same problems, and in which solutions in one subfield are equally relevant in another. Because it can channel energy toward solving a common set of problems, many geographers view GIS as a means to dissolve the long-standing regional-systematic and human-physical divisions in the discipline (Abler, 1987).

This ambitious project of building a more coherent and integrated discipline is commendable but fraught with dilemmas. To begin with, it is deeply rooted in a pragmatic vision of geography conceived as a strategy for ensuring the discipline's academic survival. Competition is intense in academia, and geography, like other disciplines, must demonstrate the value and uniqueness of its contribution if it is to outlive the current era of shrinking resources. Abler (1987) thus argues that geographers need to present a unified front, stop their "incessant whining," and discontinue the attacks that come from within the discipline. According to Abler, the business of geography is to find effective solutions to the world's problems, and this business is interfered with by the continued discussion of alternative approaches to these problems and the various "isms" existing and as yet unborn. In this pragmatic realm, difference of opinion is undesirable, for it detracts from more immediate practical needs.

Of course, preserving geography as an academic discipline is an important pursuit. Discussions of what geography ought to be—if they are indeed important—scarcely serve any purpose if geography itself does not exist. However, perhaps the problem is not that geography is too unfocused, too impractical, and too diverse, but that the academic environment has become insensitive to the nature of scientific inquiry. The attempt to re-dress geography in the cloak of applied knowledge and utilitarian values does not guarantee that solutions to the world's problems will arise as a matter of course. Many scientific discoveries of immense practical importance have arisen from research conceived without a specific application or a real-world problem in mind. Meanwhile, what percentage of articles in the geographical literature purporting to be relevant to planning or other applied fields are actually used in this capacity?

It is also not clear that geography's diversity is a flaw instead of a great strength. Geography is inherently eclectic because the discipline is defined only by a perspective on the world. However, those who advocate the computer as a means to unify geography have a particular conception of the discipline in mind, an empirical and pragmatic one that is by no means universally accepted. The numerous subfields of geography are not homogeneous in terms of the ways in which geographical reality is conceived and modeled. GIS cannot hope to address the issues of interest amid such diversity (Kellerman, 1983), since not all geographies are amenable to the model espoused by the computerized database. How would one represent geographies of the mind in a GIS data layer? Can mythical geographies be located in an absolute coordinate system? How does humanistic geography translate into objects in a spatial database? If these geographies cannot be subsumed within the realm of GIS, what is to become of them? Are they necessarily irrelevant to the geography of the future? In many subfields, the unification promised by GIS would be experienced as a radical reorientation in order to adapt to the new technological means. The effect would be a profound form of reverse adaptation.

Of course, GIS has not been universally accepted in geography and its advent has met with frequent intense criticism from within the discipline. There are those who view the technology not as a key, but rather as a threat to geography's survival. Jordan (1988), for example, sees GIS as a nonintellectual pursuit that is likely to reduce geography to a technical field without any solid disciplinary basis. Taylor (1990) characterizes GIS as a retreat into the realm of naive empiricism in which isolated facts have become more important than geographical knowledge. He sees the GIS community as a cult with a "technology-led mentality" engaging in a "high-tech trivial pursuit" that leaves geography intellectually sterile. Such statements probably typify the sentiments of many geographers. Long-standing debate over the nature and substance of geographical inquiry has been carried to a new level as advocates and detractors of GIS wage an ideological battle. Indeed it is ironic that, for all its supposed unifying potential, GIS has been so successful in intensifying the very ontological differences that underscore geography's diversity and eclecticism.

THE SOCIAL CONTEXT OF TECHNOLOGICAL INNOVATION

The tool-based model of technology serves to reinforce the notion that technologies are designed and built to fill specific human needs (e.g., Curran, 1987). According to this altruistic model, technological innovation offers salvation from a world fraught with problems, and all technologies—from the electric razor to the nuclear bomb—are tools in the service of humankind. From this vantage point, the computer's ascendance is attributable to its abilities to perform necessary tasks that cannot realistically be carried out in a manual environment. Continued advances in computer technology likewise occur because of the need for more processing power and more data handling capabilities.

What this model ignores is the importance of the social context, and especially the profit motive, lying behind much technological innovation. It also ignores the possibility that technology may invent its own need, and that technology may be built simply because the means to build it exist. In the dynamic set in motion by engineering capabilities, the question of human need is not a prerequisite for technological innovation. Rather, innovation is a goal in and of itself. Need—to the degree that it may be said to exist—is an engineered product, an artifact introduced ex post facto to rationalize continued technological innovation or to supply the foundation for commercial gain. In other instances, technology is invented solely for the purpose of servicing other technology. Without the original technology, the new technology has no purpose. Computers, for example, require software to make them solve problems, diagnostic tools to keep them functioning, tools to enhance their performance, tools to keep them running during periods of power outages and to prevent them from being destroyed by power surges, and so on.

Numerous examples come to mind in which this "technological imperative" is at work. New consumer products, for example, are introduced into the marketplace by corporations seeking to create needs where none previously existed. The introduction of these products occurs at such a frantic rate that it is indeed difficult to conceive of so many human needs waiting patiently for technological salvation. The high failure rate for these products indicates that most of them are simply not "needed" in the literal sense of the word. To take another example, in the case of nuclear weapons the question of "need" historically has appeared only as a rationalization following in the wake of technological innovation. York (1989), for example, argues that the U.S. decision to proceed with the development of the hydrogen bomb in the early 1950s resulted mainly from "technological exuberance," the desire to build the bomb simply because it was technically feasible. York suggests that this momentum has always played a strong role in the technological arms race, such that it is technology itself that prescribes the set of policies that might be considered as viable alternatives. A case in point is the massive stockpiling of U.S. nuclear weapons that occurred during the 1950s and 1960s. It was only after an enormous arsenal was in place that policymakers began a systematic examination of its strategic implications. The sheer size of the arsenal led quite naturally to notions such as "mutually assured destruction," wherein the number of weapons at one's disposal is seen as a critical factor in the ability to deter aggression.

Likewise, the factors lying behind the computer's ascendance in geography are more complex than conventional wisdom would suggest. In particular, the impacts of the computer cannot be divorced from the motives lying behind its acceptance and adoption by the geographical community. Much of the impetus fueling this growth has come from the private sector, where geographical knowledge has proved to be a practical and profitable commodity. As Waters (1989) notes, the claims made about the capabilities of computer systems are sometimes exaggerated for commercial gain. Profits, of course, are not limited to corporations directly involved in the development and marketing of hardware or software. The supply of information as a product in its own right is a quickly growing sector of the computer industry. It is clear that geographical techniques can be used in the creation of new information that has great commercial value. The development of new technical means has greatly enhanced this role and contributed to a view of information as a value-added product. Market area research, the classification of small area census data for marketing purposes, is one such example. Although characterized as simpleminded (Openshaw, 1989), this methodology has been enormously successful in enabling commercial firms to develop more efficient and profitable advertising and marketing strategies.

Enhanced symbiosis between the academic and private sectors has introduced real-world commercial motives and training for real-world careers as components of geographical research and education. These changes may lead to a more practical and relevant discipline, but at the same time the motives of commercial endeavor threaten to subsume some of the traditional noncommercial interests of the scientific community. The interests of science and commerce are often conflicting. Unlike scientific endeavor, commercial interests reflect a prevailing profit motive and a distribution of profits which, by definition, is inequitable. This inequity raises questions about the purposes of scientific endeavor, the societal groups that should benefit from it, and the motivations of scientists themselves. The commercialization of scientific research results in part from the need for academic disciplines to demonstrate the unique contributions they can provide in an era of increased competition for a shrinking pool of resources. Academic disciplines are becoming acutely aware of the fact that they can scarcely muster the resources of large commercial operations and must therefore enter into partnership with the private sector to achieve the same level of material and technical support. However, competition in the private sector may lead to a working environment that cannot support academia's traditionally free exchange of ideas and knowledge (Suzuki, 1989).

Of course, one might argue that a close relationship with the private sector offers geography an advantage over other disciplines in competing for resources. Levels of student enrollment, research funding, and professional prestige all seem to increase in proportion to the prevalence of computer technology. There is no doubt that geography could attract more than its share of these resources if it were as sexy as an MTV video, but it must be recognized that there is a cost involved. The commercialization of the discipline by those wishing merely to exploit the computer's image as a modern alternative to the filing cabinet, adding machine, or paper map ultimately undermines the scientific value of geographical research. The perceived need for enhanced computational capabilities in the absence of a concomitant rise in the sophistication of geographical theory serves to underscore the weakness of the discipline in the face of technological and commercial imperatives. It is reminiscent of how the educational potential of television has been subverted to the vastly more profitable role of purveying disposable consumer products. What is true of television is also true of the computer: There is an enormous gulf between the intellect required to invent and develop the technology and that required to program and use it.

CONCLUSION

The most profound impacts of the computer on geography occur precisely because the computer is much more than a tool. The conventional tool-based model affords the computer the status of a value-free technology, which implies in turn that technological adoption can continue unabated without fear of conflict or crisis. Since the computer is merely a tool for making existing operations more efficient, it offers no revolutionary challenge to established practices. According to this view, the computer is a means for ensuring geography's long-term viability. As a valuable tool for promoting research, education, and practical training, it is seen as the key to maintaining a competitive advantage over other disciplines in the war over funding and resource allocation. Indeed, to some, the technology is critical to the very survival of geography itself. Success in this arena is closely linked to the development of appropriate characteristics of the institutional environment that make it conducive to the incorporation of computer technology—what Batty (1988) refers to as "orgware." Factors impeding the rate and scope of technological adoption, which include a variety of financial, legal, social, and institutional constraints, are viewed rather like rocks and potholes on the road to Progress. Thus, even when purporting to go "beyond the bits and bytes," the debate about the appropriate role of computers in geography often never rises above the issue of institutional restructuring to facilitate the smooth transition from manual to automated environments (e.g., Foley, 1988).

Successful adoption of the computer, however, implies much more than the acquisition of the more-or-less tangible components of computer systems. The level of success achieved to date has not occurred simply because geography departments have managed to obtain the required hardware and software, or to build and maintain the necessary orgware. Rather, it is tied to the development of a new technical dimension within the geographer's existing realm of knowledge and expertise. It is those geographers who actively incorporate computational techniques into their research and educational activities who ultimately imbue the computer with meaning and value. Thus the most profound changes—both positive and negative—resulting from the computer arise from its ability to transform the geographer's understanding of the world and his or her means of explaining geographical phenomena. Using a computer implies more than being able to perform familiar operations more quickly and easily, and it seems unlikely that a straightforward enhancement of data-processing capabilities will be the most enduring legacy of the computer revolution.

Moreover, the impacts of the computer are neither inherently neutral nor necessarily beneficial. The computer imposes limits on thought and action as a function of its characteristic properties and the motives lying behind its acceptance and adoption. To ignore this fact is to imply that, in effect, technology develops independently of social and scientific context. The computer influences the way in which research problems are selected for study and the character of the assumptions, languages, techniques, and models brought to bear on the problem. Certain problems are more conducive to automation and sometimes the computer dictates how the problem must be structured. The insights provided by the computer are often different from those obtained by other means. Thus automation is not a simple one-to-one mapping of existing geographical techniques into a computer environment.

Despite my emphasis on negative impacts, my intent in this chapter was not to argue against the adoption of the computer in geography. Rather, my intent was to elucidate some of the implications of this adoption process that geographers have not addressed in detail in the published literature, and to stimulate debate over the appropriate role of this technology in geography. Many of the issues touched on in this chapter have received scant attention in the geographical literature. This is unfortunate since the computer revolution has had a tremendous impact on geography and will continue to reshape the discipline as computing systems become more widely accepted and adopted. While the effects of the computer revolution have certainly not been wholly positive, this does not imply that geographers should respond in the style of the Luddites, clutching axe and hammer to smash the agents of their oppression. Rather, the goal should be to gain a better understanding of these effects in order to manage this technology appropriately.

Geographers, after all, shoulder much of the responsibility for the transformations brought about by the advent and adoption of the computer in geography. These transformations must be examined with reference to the institutions within which geographers interact on a daily basis. Discussions of the effects of the computer serve no real purpose if they are simply allowed to become wrinkles on the fabric of scientific and disciplinary propriety (Winner, 1977b). Ultimately it is geographers themselves who, in the course of their daily activities, set in motion the evolutionary dynamic shaping the future character of the discipline. Carrying the discussion into the realm in which real decisions are made implies a certain responsibility shared by those who claim to be geographers, whether or not computers form an integral component of their mode of geographical inquiry.

ACKNOWLEDGMENTS

This chapter owes its existence to the patience and persistence of Dr. Helen Couclelis of the Department of Geography at the University of California at Santa Barbara. I would also like to acknowledge the contributions of the late David Simonett, whose deeply held views about the potential benefits of computer technology greatly influenced my appreciation for the complexity and importance of the issue. Of course, I alone am responsible for the views expressed, all errors and ommissions, and any misrepresentations of others' works.

REFERENCES

Abler, R. (1987). What shall we say? To whom shall we speak? *Annals of the Association of American Geographers, 77,* 511–524.

Abler, R., Adams, J. S., & Gould, P. (1971). *Spatial organization.* Englewood Cliffs, NJ: Prentice-Hall.

Batty, M. (1988). The cult of information. *Environment and Planning B, 15,* 375–382.

Blakemore, M. (1985). High or low resolution? Conflicts of accuracy, cost, quality and application in computer mapping. *Computers and Geosciences, 11,* 345–348.

Borgman, A. (1982). Technology and nature in Europe and America. In R. N. Barrett (Ed.), *International dimensions of the environmental crisis* (pp. 3–19). Boulder, CO: Westview Press.

Cowen, D. J. (1983). Automated geography and the DIDS (Decision Information Display System) experiment. *Professional Geographer, 35,* 339–340.

Cromley, R. G. (1983). Automated geography: Some problems and pitfalls. *Professional Geographer, 35,* 340–341.
Curran, P. (1987). Remote sensing methodologies and geography. *International Journal of Remote Sensing, 8,* 1255–1275.
Dawson, J. A., & Unwin, D. J. (1977). *Computing for geographers.* New York: Crave and Russak.
Degani, A. (1980). Methodological observations on the state of geocartographic analysis in the context of automated spatial information systems. In H. Freeman & G. G. Pieroni (Eds.), *Map data processing* (pp. 207–221). New York: Academic Press.
Dobson, J. E. (1983). Automated geography. *Professional Geographer, 35,* 135–143.
Foley, M. E. (1988). Beyond the bits, bytes, and black boxes: Institutional issues in successful LIS/GIS management. *GIS/LIS '88,* 608–617.
Goodchild, M. F. (1988). Stepping over the line: Technological constraints and the new cartography. *American Cartographer, 15,* 311–319.
Goodchild, M. F., & Janelle, D. G. (1988). Specialization in the structure and organization of geography. *Annals of the Association of American Geographers, 78,* 1–28.
Jordan, T. G. (1988). The intellectual core. *Newsletter of the Association of American Geographers, 23,* 1.
Kellerman, A. (1983). Automated geography: What are the real challenges? *Professional Geographer, 35,* 342–343.
Lundberg, C. G. (1989). Knowledge acquisition and expertise evaluation. *Professional Geographer, 41,* 272–283.
MacEachren, A. M., & Monmonier, M. (1992). Introduction. *Cartography and Geographic Information Systems, 19,* 197–200.
Maguire, D. J. (1989). *Computers in geography.* New York: Wiley.
Marble, D. F., & Peuquet, D. J. (1983). The computer and geography: Some methodological comments. *Professional Geographer, 35,* 343–344.
Moellering, H., & Stetzer, F. (1983). A comment on automated geography. *Professional Geographer, 35,* 345–346.
Monmonier, M. S. (1983). Comments on "automated geography." *Professional Geographer, 35,* 346–347.
National Center for Geographic Information and Analysis. (1989). *Selections from a proposal to the National Science Foundation* (Technical Report No. 88-2). Santa Barbara: National Center for Geographic Information and Analysis, University of California at Santa Barbara.
Openshaw, S. (1989). Computer modelling in human geography. In B. Macmillan (Ed.), *Remodelling geography* (pp. 70–88). Oxford: Basil Blackwell.
Pickles, J. (1991). Geography, GIS, and the surveillant society. *Papers and Proceedings of Applied Geography Conferences, 14,* 80–91.
Poiker, T. K. (1983). The shining armor of the white knight. *Professional Geographer, 35,* 348–349.
Porteous, J. D. (1986). Intimate sensing. *Area, 18,* 250–251.
Rundstrom, R. A., & Kenzer, M. S. (1989). The decline of fieldwork in human geography. *Professional Geographer, 41,* 294–303.

Slocum, T. A. (1990). The use of quantitative methods in major geography journals, 1956–1986. *Professional Geographer, 42,* 84–94.

Suzuki, D. (1989). *Inventing the future.* Toronto: Stoddart.

Taylor, P. J. (1990). Editorial comment: GKS. *Political Geography Quarterly, 9,* 211–212.

Tomlinson, R. (1989). Presidential address: Geographic information systems and geographers in the 1990s. *Canadian Geographer, 33,* 290–298.

Waters, N. (1989). Do you sincerely want to be a GIS analyst? *Operational Geographer, 7,* 30–35.

Weiser, M. (1991). The computer for the 21st century. *Scientific American, 265*(3), 94–104.

Weizenbaum, J. (1976). *Computer power and human reason: From judgement to calculation.* New York: W. H. Freeman.

Winner, L. (1977a). *Autonomous technology: Technics-out-of-control as a theme in political thought.* Cambridge, MA: MIT Press.

Winner, L. (1977b). On criticizing technology. In A. H. Teich (Ed.), *Technology and man's future* (2nd ed., pp. 354–375). New York: St. Martin's Press.

Winner, L. (1986). *The whale and the reactor: A search for limits in an age of high technology.* Chicago: University of Chicago Press.

York, H. F. (1989). *The advisors.* Stanford, CA: Stanford University Press.

CHAPTER 6

Manufacturing Metaphors

PUBLIC CARTOGRAPHY, THE MARKET, AND DEMOCRACY

Patrick H. McHaffie

What is there in this richly endowed land of ours which may be dug, or gathered, or harvested, and made part of the wealth of America and of the world, and how and where does it lie? (Congressman A.S. Hewitt of New York, author of legislation establishing the U. S. Geological Survey, 1879, in Thompson, 1981)

It may be that knowledge, in itself, is the dorsal fin of existing society, playing in public tomorrow the role which it took on yesterday somewhat more discreetly. . . . It may even be true to say that "pure" knowledge has become the axis of (technocratic) state capitalism as well as (technocratic) state socialism; it may serve as their common measure, as "real world." It may be the guarantee of change, from a society which is manipulative (of people, of needs, and of its own aims) to a society which is even more smoothly manipulative. (Lefebvre, 1973, p. 76)

The "map" has recently become the subject of a growing body of critique within geography. This critique is based in an epistemology that includes notions of deconstruction (cf. Harley, 1988, 1989, 1990; Pickles, 1991), legitimation (Harvey, 1989), and state formation (Edney, 1986). In this chapter, I extend earlier work on cartographic information as a commodity, an economic good, and the product of a specific labor process (McHaffie, 1993) by focusing on the map as a commodity in, and a product of, a mixed public–private institutional matrix. In this setting the chapter seeks to problematize the current focus on the text and opens a dialogue concerned with the "author" of the text and the nature of the cartographic labor process.

The first section highlights the creation of cartographic information within the public sector, focusing on the ideology of the capitalist state and the role of the "map" as a legitimation device in Western societies. In the second section I examine the public-sector cartographic labor process as it has been articulated within the United States. Particular attention will be paid to the entry of "scientific management" procedures in public cartographic production, the integration of technological advances such as aerial photography and photogrammetry into the labor process, and the subsequent effects these changes had upon the cartographic laborer and ultimately the map itself. This early history of technological restructuring in the cartographic labor process is juxtaposed against recent changes in the production of digital cartographic information and its utility in the emerging information economy, becoming what David Rhind (1992) calls "the elixir of life" for GIS. In the final section I will describe the marketing of this public good within the context of a major federal mapping agency in the United States and reflect on its implications for a more-or-less democratic cartography.

My purposes here are twofold. The emergence of new communications technologies has created a climate where a conflation of access to cartographic information and democracy has occurred in the disciplinary mind of many geographers and geographic information systems (GIS) specialists. The potential of this developing technology has created a false sense of egalitarianism among champions of the "new" cartography and GIS. I hope to recast this issue into a more realistic mold, exposing the commodified nature of digital cartographic information. In addition, I hope to open a discourse that contextualizes the actual human activities surrounding the production of public cartographic information, reconnecting the "cartographer/ author" to the "map/ text" at the point of production and portraying cartographic information as the product of an industrial process, while at the same time offering a sympathetic alternative to the "map as text" metaphor.

THE CREATION OF PUBLIC CARTOGRAPHIC INFORMATION

The social investment in the creation of large-scale "general purpose" mapping systems has become an essential element in the systematic public provision of infrastructure. The built environment (buildings, roads, public infrastructure, etc.) is made "real" before its actual construction through the inclusion of its components in urban plans, regional and national transportation plans, and other planning maps that are seized upon by the public as "maps" of the future. Chambers of commerce and local development agencies vie for that particular combination of infrastructure, convention facilities, corporate headquarters, or state investment that will "put us on the map." So the act of "putting something on the map" or conversely "leaving something off the map" becomes at once a critical moment in the operation of the space economy and a legitimation of the relations of production required to produce space, painted by Harley (1990) as "cartographic complicity":

> To map the name was to give its prejudice an anonymous legibility. To publish the name was not only to make it permanent but also to give it authority and legitimacy as a coordinate on a federal map. (p. 4)

Thus, the publication of placenames that contain racial and ethnic bias is merely an example of the wider "microphysics of cartographic power" (Harley, 1990, p. 3). Space as commodity, and as a necessary resource in the survival of capitalism, becomes codified reality under the Cartesian gaze of federal mapping programs. Harvey (1989) characterized the role of these various projects in *The Condition of Postmodernity:*

> The conquest and control of space, for example, first requires that it be conceived of as something usable, malleable, and therefore capable of domination through human action. Perspectivism and mathematical mapping did this by conceiving of space as something abstract, homogeneous, and universal in its qualities, a framework of thought and action which was stable and knowable. (p. 254)

The U.S. Public Land Survey of the western United States, with its rational grid of 36-square-mile townships, was part of a process of land privatization/commodification and was embedded within a utilitarian project of fragmentation and control, separating indigenous peoples from their ancestral territorial base and claiming the newly created spaces of interior America for the expanding federal state. This embodiment of the modern project of enclosure, control, and compartmentalization of spaces, societies, and territories has been extended in the 20th century with the promulgation of modern topographic base maps. By basing the subdivision of space on a worldwide grid such as latitude and longitude, these mapping systems tear local meaning from areas included within the margins of the "quadrangle," placing cities and towns, roads and railroads, mines and mills, ghettos and subdivisions, indeed all of the built environment, into a global context. This subdivision of the world is becoming even more institutionalized as digital cartographic databases are created which are linked to the rational divisions instituted by modern topographic mapping systems.

The subdivision of lived space, its role in the functioning of the capitalist economy, and the analysis of this activity has become a high priority in human geography. The creation of a national mapping policy (hinted at in the opening quotation) began in the United States at an early stage in its history, first spurring the mapping of the coasts to ensure the safety of early commerce, and later driving the mapping of the vast natural resources of the interior. The mapping of these spaces commodified them and the natural resources they contained; it "created" space as an exploitable resource. According to Lefebvre (1973, in a chapter appropriately titled "The Discovery"),

> what has happened is that capitalism has found itself able to attenuate (if not resolve) its internal contradictions for a century, and consequently, in the hundred years since the writing of Capital, it has succeeded in achieving "growth." We cannot calculate at what price, but we do know the means: by occupying space, by producing a space. (p. 21)

The mapping of such "produced spaces" became an essential element in their production. In the United States, the state apparatus responsible for this mapping was the United States Geological Survey (USGS).

THE CARTOGRAPHIC LABOR PROCESS AS A STATE ENTITY

Cartographers and cartographic historians have begun to view more critically the act of producing representations of spatial reality (Harley, 1989; Edney, 1986). Harley's (1989) deconstruction of cartographic objectivity linked cartographic process to larger societal structures of power and exploitation:

> Power comes from the map and it traverses the way maps are made. The key to this internal power is thus cartographic process. By this

> I mean the way maps are compiled and the categories of information selected; the way they are generalized, a set of rules for the abstraction of the landscape; the way the elements in the landscape are formed into hierarchies; and the way various rhetorical styles that also reproduce power are employed to represent the landscape. To catalog the world is to appropriate it, so that all these technical processes represent acts of control over its image which extend beyond the professed uses of cartography. The world is disciplined. The world is normalized. (p. 13)

Viewing maps as embodiments of power relations directly challenges the traditional scientific, technical notions of objectivity that permeate cartographic practice. But if the key to the internal power of maps is the "cartographic process," then the sociology of that process must be analyzed in order to understand how power flows through it (Rueschemeyer, 1986; Grint, 1991).

During the late 19th and early 20th centuries industrial capitalism was able, to a degree, to resolve crises in Western Europe and North America through the implementation of various configurations of Taylorist and Fordist "scientific management" schemes (Gartman, 1979; Littler, 1982). These revolutionized the industrial production process and restructured industrial production into fragmented but linked serial processes. A key element of Taylorist formulations is the physical and social separation of the conception of the task and its execution, as well as the reduction of the production process to a number of more simple connected tasks (Taylor, 1911; Braverman, 1974). This type of production is commonly known as "production line" or "assembly line" production.

Not only is the production process divided, however, but the social relations of production are fragmented into horizontal and vertically integrated "cells." It becomes more difficult for a production worker to understand just who he or she is working for and where and how his or her contribution to the process fits into "the big picture." The separation of conception and execution, hand and brain, opens the door to the development of antagonistic social relations between managers and workers. As Braverman (1974) pointed out:

> A necessary consequence of the separation of conception and execution is that the labor process is now divided between separate sites and separate bodies of workers. In one location, the physical processes of production are executed. In another location, are concentrated the design, planning, calculation, and record-keeping. The pre-conception of the process before it is set in motion, the visuali-

> zation of each worker's activities before they have actually begun, the definition of each function along with the manner of its performance and the time it will consume, the control and checking of the ongoing process once it is under way, and the assessment of results upon the completion of each stage of the process—all of these aspects of production have been removed from the shop floor to the management office. (p. 124)

The first major differentiation of cartographic laborers occurred with the invention and adoption of the printing press, when compilation and reproduction of the map were separated (Monmonier 1985, pp. 145–146). Mass reproduction commodified the map image and created a distinctive cartographic labor process. The map became one of a growing list of "objects" deemed to be necessary to rational society and the expanding space economies of Western Europe. The map's role in consolidating state power and the need for military secrecy ensured that maps would be produced within the state apparatus. Private-sector mappers developed expertise in small-scale, highly generalized representations or local area maps with limited local markets, while the production of expensive large-scale "general purpose" (topographic and geodetic) cartographic systems was left to the state. Whether the "purpose" of these maps was general use (i.e., use by a large segment of the public) or general application (i.e., use by a specific portion of the public for specific purposes in any area) depended on the interpretation of the cartographic historian.

In the 20th century the U.S. federal government developed large-scale public mapping systems as the necessary topographic precursor to geological, hydrological, and botanical investigations west of the Mississippi River. The federal commitment to this policy was created within the context of bureaucratic/institutional tension between the federal and state governments regarding responsibility for the funding of scientific endeavors within their respective borders (Edney, 1986). The development of a national mapping policy and the establishment and expansion of the two major 20th-century civilian mapping agencies, the U.S. Coast and Geodetic Survey and the U.S. Geological Survey (USGS), came as the result of ad hoc legislation designed to delineate responsibilities between agencies at the federal and state level. This policy was constantly under attack during the late 20th century, largely due to its expense and the acerbic view held by many interior state legislators of the federal role in funding scientific activities. The piecemeal nature of the legislation dealing with responsibility for surveying and mapping resulted in a tension between state and federal mapping programs, so that no coherent federal policy was legislated until the Temple Act in 1925. This act committed the federal government to the completion of a "general utility topographical survey" of the entire country within 20 years. Once the federal government was committed to large-scale mapping, the stage was set for the transformation of the public cartographic labor process in the United States.

The transformation of the public cartographic labor process in the United States displays elements of a corporatist logic that borrowed the most efficient components of developing industrial labor (Taylorist) processes in the late 19th and early 20th centuries. The resulting system would more closely resemble a modern factory than the craft/apprentice organization that had predominated prior to World War I (McHaffie, 1993). Before World War I, most of the labor involved in topographic mapping was accomplished in the field. The tradition and myth of the mapper as a rugged surveyor single-handedly mapping and taming the vast Western wilderness was created during this period. However, after World War I, many USGS topographers returned from military service with experience gained in the use of aerial photography in reconnaissance and mapping. They "applied their interest in aerial photography to its potential use in the civilian topographic mapping program. Throughout the 1920s they experimented with applications of the relatively new science of photogrammetry and succeeded in making a few maps from aerial photographs" (Thompson, 1981, p. 7).

The incorporation of aerial photography into the cartographic labor process allowed the restructuring of the relations of production within the USGS to a more efficient, more easily controlled Taylorist formulation. The aerial photographer could, in a single day, photograph hundreds of square miles and supply technicians with the necessary materials for map compilation. The mappers were no longer required to "slog" into the messy reality of the field in order to produce the "map"; they no longer were required to compile their manuscripts in the field. In fact a "new" cartographer was created, one laboring in dark rooms using complicated optical-mechanical instruments, embedded within a process that was more easily watched, more easily controlled. Within agencies such as the USGS and the Defense Mapping Agency (DMA) this process has since been refined and the technology has since been modernized through improvements in aerial platforms, cameras, film, optics, and electronic digital storage and processing, but the conditions of work for laborers in this stage of cartographic production have remained fundamentally the same. The need for field survey and the actual contact of the cartographer with the object of his or her work was, as a consequence, greatly reduced. A new technique of aerial photograph interpretation was born, elaborated, codified, and institutionalized within the state.

The labor process is the instrument that shapes the map as an embodiment of larger societal power relations. Cartographers serve not merely as channels of existing power relations, but also act independently within the constraints of a particular production process to constantly redefine and reshape the cartographic messages that pass from the state to the public. The specificity of the labor process, as constituted within the state apparatus, determines the conditions under which public cartographic information is produced. But the public cartographic labor process is peculiar as a state formation in that it assumes a Taylorist configuration within the state *in the absence of the logic of accumulation that is assumed to drive similar schemes in the private sector.* This situation is comparable to other industrial production processes that have been promulgated under the direction of the state, notably the production of actual specie within government mints and the production of documents by government printing offices. However, both of these processes are merely concerned with the reproduction of a preformed message. In the production of cartographic information the cartographer is, to a greater or lesser degree, the author of the created information.

The history of the cartographic labor process has been characterized by the division of labor into a series of disconnected operations, with each operation contributing to the final message embodied within the map. This history has been one of fragmentation, rationalization of production procedures, and progressive deskilling of the cartographic laborer. If "cartographers manufacture power" as Harley (1989) states, then public-sector cartographers do so within circumstances that are shaped by the particular institutional configuration of the state cartographic production process. This configuration can take many forms, from the one-person shop in the smallest local government office producing representations using pen-and-ink techniques, to huge agencies such as the DMA responsible for producing global mapping systems in a significantly different technological environment.

As the cartographic production process has been progressively fragmented and rationalized, the cartographic laborer has become increasingly alienated from the product of his or her labor: cartographic information. Each step is geared toward the production of particular "map separates." These layers of information are then combined to produce a composite image. However, the conception of the process, of how these particular separates will fit together to produce the map image, has been gradually removed from the domain of the detail worker.

Figure 6.1 shows a schematic diagram of the cartographic production process as conceived and represented by the USGS (Thompson, 1981, p. 24). Production is organized as a linear, one-way process consisting of a number of tasks along the way. These tasks are particular moments in the production process, sites of transformation of the incipient map into something closer to the "final" product. They are

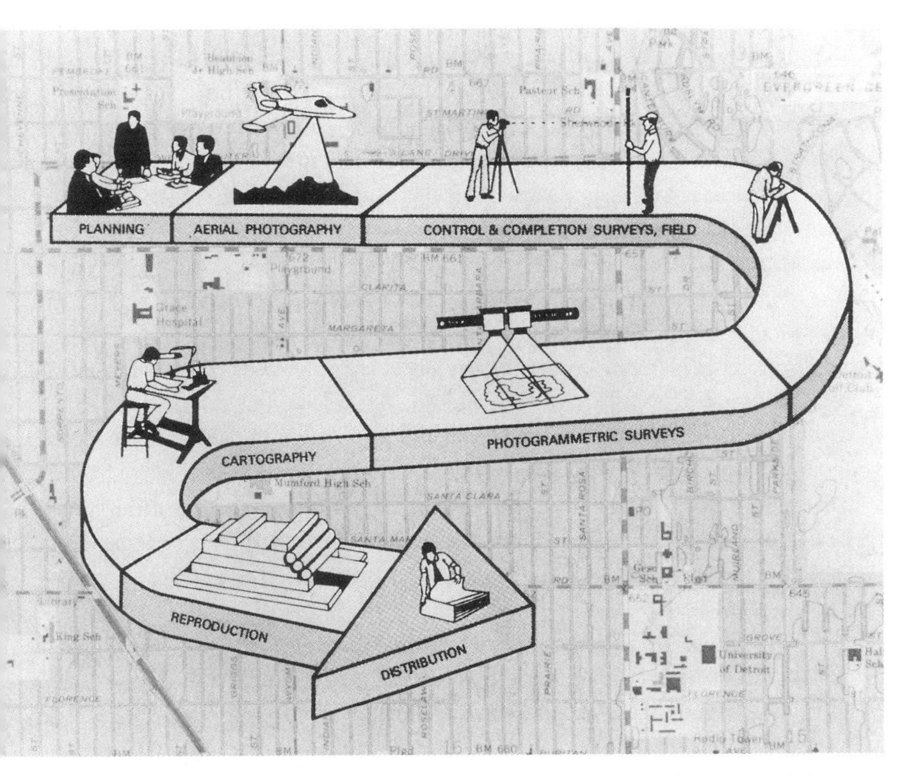

FIGURE 6.1. The cartographic product process. From Thompson (1981, p. 24).

sites of the creation of public cartographic information. Each station along the line is only a part of the labor process, connected through the production sequence to two other parts of the production process: that which immediately transformed the map before it arrived in your shop, and that which requires the transformed product you will produce. It is significant that the process is schematically represented as a conveyor belt, a machine used for moving product from fixed site to fixed site inexorably closer to the point of contact with the user, "Distribution." The seven sites are alternately peopled (manned?) by workers or by machines. Integration of technology and human labor progressively devalues the human contribution to the map. Workers are increasingly viewed as "technicians," "operators," or "warm bodies." At this point, the job category of "topographer," the maker of the topographic map, has disappeared and has been replaced by specialists. This view of the cartographic laborer reflects a historical deskilling of production workers that has proceeded hand in hand with the fragmentation of the labor process achieved under "scientific management" schemes.

The incorporation of aerial photography and photogrammetric techniques into public-sector mapping has been the most transformative of all technological developments in the cartographic labor process. It has, to a large extent, created and furthered the aura of scientific objectivity that enshrouds topographic mapping today. If the cartographer is physically separated through the abstraction of the photograph from the object of his or her attention, then the resulting product, the topographic map, will somehow be objective, value-free, and scientific. This assumption of objectivity can create an ambivalence toward the object of knowledge. Perhaps the "frightened Africans" who once "threw spears at an Aero Service aircraft" or the "suspicious moonshiners in Appalachia" who "took a few rifle shots" at aerial mappers did so not because the intentions of the mappers were "not always understood," but because those intentions, and the powerful forces behind them, were understood only too well (Wilford, 1981, pp. 236–237).

The pioneers of photogrammetry—Fairchild, Abrams, Goddard, Kauffman, and Brock—sold the new technology of aerial mapping to the USGS based on the improvements in accuracy that it produced (Heiman, 1972). This, and other new technologies in the area of photographic processes and reproduction, further increased the need for linked systems of specialists. They took on new names (e.g., operators, scribers, photographers, photogrammetrists), assumed newly deskilled roles, and conformed to the Taylorist fragmentation of the labor process in a corporatist spirit of increased productivity, loyalty, and efficiency. The "scientific management" project of control, projected into the cartographic labor process, was reflected back onto society in the form of rationalized and increasingly depersonalized representations of the modern world. To witness the transformation of cartographic representation as technology is introduced one needs only study, for a particular area, the 30-minute maps produced by the USGS in the 1880s, the 15-minute maps produced by the USGS in the 1920s, the 7.5-minute maps produced by the USGS in the 1950s, and the DLG's (Digital Line Graphs) produced by the USGS in the 1980s. Each successive map series presents a more standardized, more accurate, more codified product. The "mistakes" or artifacts produced by a closer human involvement disappear.

The technological revolution in cartography since the 1960s has not yet produced a rethinking of the major mapping agencies' role as providers of cheap, subsidized digital cartographic information. But in 1990 a change in focus was signaled with the completion by the USGS of the primary quadrangle map series (7.5 minute). According to Ramey (1992, p. 1114), "During the 1990's emphasis will shift to adding data to the National Digital Cartographic Data Base (NDCDB) and to digital revision as digital data become available and advanced digital techniques are developed and refined. . . . The primary objective is to create a current and comprehensive database from which the USGS can quickly deliver up-to-date digital data and digitally derived map graphics to meet the needs of the user community." The first step in accomplishing this task will be to "expand and improve mass digitization capabilities," although it is assumed that this will not be done through the creation of state-sponsored "electronic sweatshops" (Garson, 1988). Other steps include the improvement of information content, revision and production capabilities, and quality control. The message, however, is clear. The primary federal mapping agency has seen the future and the future is digital. In the new economy the Digital Elevation Model (DEM) and the DLG are the new currency. The creation of standards for digital data such as the Spatial Data Transfer Standard (SDTS) under the guiding hand of the USGS in effect codifies that economy and ensures the agency's bureaucratic ascendancy. Whether the new digital cartographic production system will markedly change the conditions of work for the cartographic laborer remains to be seen. The establishment of priorities within the primary mapping agency signals a continuation of the philosophy of mass production and distribution, and the rhetoric of official publications reflects a constant sense of struggle to meet user "needs" and "demand." Under this pressure to produce while struggling to overcome bureaucratic inertia, it is likely that the new system will mimic the historical success of the state-Fordist "map factories" that revolutionized cartographic production in the postwar period, albeit within an evolving technological configuration.

CARTOGRAPHIC INFORMATION + THE MARKET = DEMOCRACY?

In the National Mapping Program of the USGS, the logic of the market has found its way into the production and distribution of digital products through several channels. Information regarding products in demand is provided to the agency through state mapping advisory committees (SMACs, normally composed of industry and government map users), through cartographic information clearinghouses such as the Earth Science Information Center (ESIC, formerly the National Cartographic Information Center), and through the direct monitoring of product sales. This information then guides the production of new "product" for a supposedly "general" market. Marketing logic enters into the practices of public distribution offices in the form of pricing policies, advertising, and promotions, but in the absence of the profit motive guiding private-sector enterprises. Success or failure of the distribution office is measured in terms of numbers of maps sold or customers serviced. Drops in sales or requests for cartographic information may result in personnel cuts or reduced funding for the sales office. These slumps in sales can also be viewed as an incentive for public agencies to better advertise their wares. The logic of "marketing" and its effect on the more widespread adoption of new map technology is outlined by Lauer, Estes, Jensen, and Greenlee (1991):

> For a data set to be available to potential users, the original program requires a *marketing* effort that includes a *distribution* mechanism and an *institutional commitment* to furnish the data in formats and on media that may not have been required by the original program. . . . It requires an *institutional willingness* to *consider the needs of users and to modify standard products to meet those needs.*

> These service efforts require resources that are usually not included in the original program. The required resources may represent only marginal increases in the base program, yet the *investment* is not made. Only the largest programs in digital cartography, such as those conducted by the USGS and the Bureau of Census currently provide an adequate *product awareness activity.* (p. 648; emphasis added)

The development and production of new cartographic products (in this instance given in reference to the USGS cartographic and remote sensing data distribution system as well as the U.S. Census Bureau's TIGER data distribution system) are assumed to be guided by the "invisible hand" of user demand. Furthermore, the producer must, in advance, consider the needs of users and modify products to meet those needs. But just who are the "users" and does the invisible hand always know best?

The Landsat program can serve as an illustration of the inherent problems of mixing a "free market" ideology with subsidized information production. The program was established as an experimental system for gathering repetitive worldwide multispectral data for civilian uses.[1] Through the 1970s and into the 1980s the Landsat program was managed by the USGS through the EROS Data Center and provided low-cost photographic products and digital data to corporations, government agencies, educators, students, researchers, and foreign governments. In the 1970s satellite photographic images were priced from $8 to $50, while digital scenes were sold for $200. Pricing under this regime was limited to the cost of reproduction, rather than the high cost of acquiring the data. The cost of system development and maintenance was subsidized through annual appropriations from Congress. During the 1980s, however, the program was first transferred to the National Oceanic and Atmospheric Administration (NOAA) and then it was finally privatized under the Earth Observation Satellite (EOSAT) Company. Price increases during the NOAA transition and after commercialization were intended eventually to pass on the full cost of acquiring the data and maintaining the system to the consumer. During the period between 1980 and 1990 the cost of photographic products, arguably the most "democratic" product (requiring the least capital investment to use the image), has increased as much as 50 times for some comparable items. During the same period the prices of digital products were increased up to five times their precommercial cost. In addition, copyright restrictions were placed on the use of both photographic and digital products, effectively removing all imagery produced after 1985 from the public domain and severely limiting their use. During the same period several other nations have launched successful private-venture and state-subsidized remote sensing satellites. One result of these developments has been a precipitous drop in the number of photographic products sold. In 1980, at an average price of $15, there were over 128,000 Landsat film products sold. This compares to just over 4,000 sold in 1989 at an average price of $150. Digital products have seen a slow but steady increase, still amounting to only 7,374 computer-compatible tapes sold in 1989. Who are the consumers?

> Landsat data are now being purchased primarily by only a few government agencies and a number of aggressive corporations. Research facilities, academic institutions, educators, students, State and local governments, and the governments of less developed nations are now purchasing considerably less data than they did a few years ago. (Lauer et al., 1991, p. 649)

The USGS has also aggressively moved toward the development of new formats, new products, and new standards for the rapidly evolving NDCDB. The traditionally "gray" *USGS Yearbook for Fiscal Year 1989* announced the development of a new product, the "Digital Topographic Map of the Future," a set of county-based digital data sets on CD-ROM. The giddy announcement reads much like standard advertising copy, hardly what one would expect from the primary mapping agency:

> This project proposes to test not only a *novel medium* for data distribution (CD-ROM) but also *novel spatial coverage* (county versus the traditional quadrangle base) and a *novel marketing approach* in which data display and processing software will be packaged with the data. (U.S. Geological Survey, 1990, p. 65; emphasis added)

Indeed, the CD-ROM is a relatively new medium for marketing a high volume of digital data in an attractive and inexpensive format. But the coverage of county areas as opposed to quadrangles is hardly an innovation. In fact, county maps and atlases predated most of the modern quadrangle map coverage by several decades and most states still produce county-format maps in some fashion. During the early 1980s the USGS, in response to user demand, had begun producing paper county topographic maps. The paper or tactual map has always contained the inherent ability to be used by the reader with a minimal amount of peripheral aid or equipment.

The economist J. M. Keynes said, "Capitalism is the extraordinary belief that the nastiest of men for the nastiest of motives will somehow work for the benefit of us all" (quoted in Sherrill, 1984, p. 355). Whether one agrees with this rather dour view of the market economy and its workings or not, even the most ardent supporters of free enterprise must admit that the benefits of laissez-faire capitalism are unevenly distributed. The "belief" cited above is a good example of an *ideology:* "an inverted, truncated, distorted reflection of reality" (a paraphrase of Jorge Larrain, by Smith, 1984, p. 15). If the production of cartographic information is organized and controlled with reference to this ideology, albeit in a mixed public/private institutional matrix, and within a scientific/technological milieu that is in a state of constant transformation, it remains to be seen how this economy will interact or draw sustenance from another enduring ideology, that of democracy.

Many have claimed that we are on the verge of a new era of *electronic democracy.* Jim Warren (1993), writing in *Government Technology* under the heading "Electronic Democracy: The Year of the Internet (Networking Hits Critical Mass)," refers to electronic mail, international forums or discussion lists of "experts," computer conferencing, and the proliferation of on-line data as a signal that new forms of access to information will result in a more democratic society.[2] Indeed, the networking revolution and the growth of electronic networks such as Internet, Bitnet, and Fidonet has reconfigured the technology of information distribution and created new avenues for "grass-roots" initiatives to become more generally distributed. The White House came "on-line" on June 1, 1993, and is providing texts of addresses and position papers at a staggering rate.[3] Perhaps those of us with "access" have become jaded and forgotten that the 14.2 million individuals claimed to be connected in some way to one of the international computer networks represent only a small fraction of the 5 billion or so individuals on the planet (substantially less than 1%). Cyberpunk author Neal Stephenson (1993), in his novel *Snow Crash,* projects a bleak future for electronic democracy:

> In the real world—planet Earth, Reality—there are somewhere between six and ten billion people. At any given time, most of them are making mud bricks or field stripping their AK-47s. Perhaps a bil-

> lion of them have enough money to own a computer; these people have more money than all of the others put together. Of these billion potential computer owners, maybe a quarter of them actually bother to own computers, and a quarter of these have machines that are powerful enough to handle the Street [*cyberspace*] protocol. That makes for about sixty million people who can be on the Street at any given time. (p. 26)

For billions the possibility of accessing the best technology and information made available through digital communications networks will always be a luxury. Cartographic information, digital or otherwise, becomes a commodity in its mass production and marketing. If market forces are guiding the production of this commodity, even though it is portrayed as a "general purpose" product, then it will be produced to serve a particular constituency, in particular those who serve to gain from the "digging," "gathering," or "harvesting" the wealth of America and of the world.

NOTES

1. For a detailed account of the Landsat system, its history and capabilities, as well as some of its applications, see Avery and Berlin's (1992) *Fundamentals of Remote Sensing and Image Interpretation,* pp. 134–147.

2. The article by Warren (1993), first in the new series in the tabloid, is placed, not by coincidence, in the annual GIS issue. GIS'ers can be prolific Internet junkies, as anyone who has subscribed to the discussion list GIS-L out of the University of Buffalo will attest.

3. CompuServe Mail can be sent to the White House at the following CompuServe Mail addresses: CIS: President and CIS: Vice-President. Messages sent through the internet should be sent to the following addresses: president@whitehouse.gov and vice.president@whitehouse.gov.

REFERENCES

Avery, T. E., & Berlin, G. L. (1992). *Fundamentals of remote sensing and image interpretation.* New York: Macmillan.

Braverman, H. (1974). *Labor and monopoly capital: The degradation of work in the 20th century.* New York: Monthly Review Press.

Edney, M. (1986). Politics, science and government mapping policy in the United States, 1800–1925. *American Cartographer, 13*(4), 295–306.

Garson, B. (1988). *The electronic sweatshop: How computers are transforming the office of the future into the factory of the past.* New York: Simon and Schuster.

Gartman, D. (1979). Origins of the assembly line and capitalist control of work at ford. In A. S. Zimbalist (Ed.), *Case studies on the labor process* (pp. 193–205). New York: Monthly Review Press.

Grint, K. (1991). *The sociology of work: An introduction.* Cambridge, England: Polity Press.

Harley, J. B. (1988). Maps, knowledge, and power. In D. Cosgrove & S. Daniels (Eds.), *The iconography of landscape* (pp. 277–318). Cambridge, England: Cambridge University Press.

Harley, J. B. (1989). Deconstructing the map. *Cartographica, 26*(2), 1–20.

Harley, J. B. (1990). Cartography, ethics and social theory. *Cartographica, 27*(2), 1–23.

Harvey, D. (1989). *The condition of postmodernity.* Oxford: Basil Blackwell.

Heiman, G. (1972). *Aerial photography: The story of aerial mapping and reconnaissance.* New York: Macmillan.

Lauer, D. T., Estes, J. E., Jensen, J. R., & Greenlee, D. D. (1991). Institutional issues affecting the integration and use of remotely sensed data and geographic information systems. *Photogrammetric Engineering & Remote Sensing, 57*(6), 647–654.

LeFebvre, H. (1976). *The survival of capitalism: Reproduction of the relations of production* (F. Bryant, Trans.). London: Macmillan.

Littler, C. R. (1982). *The development of the labour process in capitalist societies.* London: Heinemann.

McHaffie, P. (1993). The public cartographic labor process in the United States: Rationalization then and now. *Cartographica, 30*(1), 55–60.

Monmonier, M. (1985). *Technological transition in cartography.* Madison: University of Wisconsin Press.

Pickles, J. (1991). Texts, hermeneutics and propaganda maps. In T. Barnes & J. Duncan (Eds.), *Writing worlds: Text, metaphor and discourse* (pp. 193–230). London: Routledge.

Ramey, B. S. (1992). U.S. Geological Survey National Mapping Program: Digital mapmaking procedures for the 1990's. *Photogrammetric Engineering & Remote Sensing, 58*(8), 1113–1116.

Rhind, D. (1992, November). War and peace: GIS data as a commodity. *GIS World, 5*(11), 37–39.

Rueschemeyer, D. (1986). *Power and the division of labour.* Stanford, CA: Stanford University Press.

Sherrill, R. (1984). *Why they call it politics: A guide to America's government.* New York: Harcourt Brace Jovanovich.

Smith, N. (1984). *Uneven development: Nature, capital, and the production of space.* Oxford: Basil Blackwell.

Stephenson, N. (1993). *Snow crash.* New York: Bantam Books.

Taylor, F. W. (1911). *The principles of scientific management.* New York: W. W. Norton.

Thompson, M. M. (1981). *Maps for America* (2nd ed.). Reston, VA: U.S. Geographical Survey.

U.S. Geological Survey, (1990). *United States Geological Survey yearbook, fiscal year 1989.* Washington DC: U.S. Government Printing Office.

Warren, J. (1993). Electronic democracy: The year of the Internet. *Government Technology, 6*(7), 22.

Wilford, J. N. (1981). *The mapmakers.* New York: Random House.

CHAPTER 7

Marketing the New Marketing

THE *STRATEGIC* DISCOURSE OF GEODEMOGRAPHIC INFORMATION SYSTEMS

Jon Goss

Marketing has become the center or "soul" of the corporation. (Gilles Deleuze, 1992, p. 6)

The cult of information is nowhere more deeply entrenched than on this bustling, highly competitive, entrepreneurial frontier, where many of the brightest young minds of the time are hard at work merchandising the promise of the computer, cleverly maneuvering its services into every slightest need that an eager, often gullible business community might present. (Roszak, 1986, p. 182)

This chapter is concerned with the integration of geographical information systems (GIS) with geodemographic databases in marketing research. This involves the combination of sophisticated computerized mapping, and spatial and statistical analysis, with proprietary consumer databases. Although the terms "market mapping" (Baker & Baker, 1993, p. 9) and "geodemographic research systems" (Curry, 1993, p. 199), have been suggested for this innovation, I shall here refer to it as geodemographic information systems (GDIS).

GDIS is a rapidly growing frontier of information technology—it is the fastest growing segment of the GIS business (GIS World, 1993)—characterized by intense competition between many relatively small, innovative corporations, but it is rapidly consolidating, via mergers, acquisitions, and licensing agreements between GIS vendors, database managers, and list brokers, into a number of "information conglomerates" (Curry, 1993, p. 27). The main players now offer modular, PC-based decision-support systems fully integrated with massive proprietary databases composed of both individual and aggregate information consolidated from public and private sources, including the following: government records such as property registration, vehicle and driver licensing, police crime reports, birth and death certifications, and the U.S. Bureau of the Census; records of private institutions such as banks, insurance companies, and health services provided in compliance with legislation; marketing customer information file (MCFI) systems which compile commercial information on consumers from credit bureau reports, customer mailing lists, and consumer surveys; and consumer studies such as panel surveys, syndicated media reports, and life-style and psychographic segmentation profiles.

The client's own customer databases and geography can be integrated with these systems to identify "prospects" (other potential customers), or to optimize retail allocation decisions such as store location and sales force assignation. For a relatively modest investment, GDIS promises to "unleash marketing power," allowing the corporate executive to turn a "marketing office into a data powerhouse with maps and reports on consumers and businesses anywhere in America" (Holtz, 1992, p. 129).

One may, of course, dismiss such claims as hype, but marketing professionals seem to believe that the systems are able to predict consumer behavior sufficiently accurately to be profitable (Baker & Baker, 1993; Holtz, 1992; Curry, 1993; Hughes, 1991), and the phenomenal growth of the GDIS industry, as well as sales of GDIS systems (Thomas & Kirchner, 1991), is surely a testament to the commitment of corporate executives to the "new marketing research systems" (Curry, 1993). And if a part of this commitment is simply due to their "gullibility" to the clever merchandising of the computer's promise (Roszak, 1986, p. 182), this begs the question of why this is the case. What is it that is so attractive about GDIS, and what do marketers think they are getting?

I will argue that the corporate community is eager to purchase GDIS because of its capacity to monitor, model, and control consumer behavior, and ultimately because they promise the capability to manipulate the market and consumer identity to enhance profitability.

Executives are eager to buy (into) a particular representation of social reality in GDIS discourse—that is, a rationally ordered, predictable, and controllable, or "productivist" (de Certeau, 1984, p. 8), world of consumption. In this world not only do producers market identities to consumers through commodities, they also market the identities of consumers as commodities. This conceit promises that consumption will become merely another moment in production and ultimately that consumers will become commodified inputs in a "total" production process.

The evidence for my argument comes from the professional literature of GDIS, including handbooks, informational brochures, and journal advertisements. I argue that the discourse of GDIS is organized around a set of metaphors consistent with a "strategic" conception of reality and identity. Although several of the organizational metaphors are secondary, that is, they are derived from other discourses (such as information technology and marketing), I argue that they are in turn organized and unified by the structural metaphor of *strategy* in the sense of that term developed by Michel de Certeau (1984). Before beginning this analysis, however, I will discuss the two main components of GDIS, that is, geodemographics and GIS.

GEODEMOGRAPHICS

> *Virtually every company—from Sun Microsystems Inc. to Sun Septic Systems—is obsessed with knowing everything about its customers, from what they eat to what they drive and what they read, so they can launch promotional campaigns at new customers who fit a given profile. (Roberts, 1992, p. 26)*

Many observers, whether critical social scientists (Leiss, Klein, & Jhally 1986), futurists (Toffler, 1980; Naisbitt & Aburdene, 1990), or marketing specialists (Baker & Baker, 1993; Hughes, 1991; Holtz, 1992), have observed the changing nature of the contemporary consumer market and the resulting shift from the mass marketing of homogeneous commodities to niche marketing or micromarketing of highly differentiated products. The transition is undoubtedly connected with the restructuring of capitalism, and attendant social, political, and cultural changes, that is, the generalized demise of the organized or Fordist regime of production, and the development of "consumer capitalism"

(Baudrillard, 1981), "disorganized capitalism" (Lash & Urry, 1987), or "flexible accumulation" (Harvey, 1989). Or, in the language of the information industry itself:

> Slower population and category growth, intensifying competition from around the globe, cautious consumers, a splintered marketplace and increasingly demanding retailers are fundamentally changing the way you do business. (National Demographics & Lifestyles, 1992a)

However, product differentiation and the targeting of specific consumers is not a recent phenomenon, and is in fact inherent in the modern capitalist circulation of commodities. Product differentiation, for example, occurred shortly after the birth of the archetypical form of modern capitalist production—in the "Sloanism" of the auto industry of the 1920s (Slater, 1992, p. 202). Selective lists of consumers and their habits have also been systematically maintained and rented out to retailers at least since R. L. Polk began assembling its consumer database in the late 19th century, and again in the automobile industry since at least 1922 when R. R. Donnelley and Sons began compiling lists of automobile and truck owners. Indeed, as Veblen (1899/1954) famously realized, marketing depends upon defining differences between commodities that resonate with real or imagined differences between consumers. Marketers have always sought to differentiate products and segment consumers for this purpose (see also Fullerton, 1985). Effective target marketing has only been limited by the development of the quality and quantity of information on consumers, the conceptual and technical means by which to analyze it, and the instruments through which to exploit it.

The development of a sophisticated target marketing system was not practicable without two social and technological innovations: the development of large and fast computers with their capacity for database management and multivariate statistical analysis; and the institutionalization of the U.S. Postal Services Zone Improvement Plan (ZIP), which allowed marketers to delineate trade areas and potential markets, and, most importantly, to reach consumers cheaply and quickly through bulk mailings (Curry, 1993, p. 201).

In the early 1970s, Jonathan Robbins, a sociologist turned entrepreneur who is conventionally credited with the invention of geodemographics (Burnham, 1983, p. 90; Weiss, 1988, p. xii), combined the theories of the Chicago school, particularly the notion of ecological competition for urban space between social groups, with the related "number-crunching" factorial ecologies of positivist urban social sciences, to produce geodemographic profiles of residential ZIP code areas for the entire United States. This method, which he called PRIZM (Potential Rating Index for ZIP Markets), involved a cluster analysis (geodemographic segmentation) of aggregate data derived from the U.S. Bureau of the Census on 240,000 block groups and individual data from consumer surveys, to produce 40 life-style clusters describing all 36,000 ZIP codes.

The cluster model is "based on the fundamental sociological truism that 'birds of a feather flock together' " (Claritas/National Planning Data Corporation, n.d.), or that "You are where you live."[1] The precise algorithms are proprietary, but clusters are typically derived through a two-step multivariate statistical process: first, a factor analysis of demographic data available for given geographic units; and second, a cluster analysis of the factor scores in each unit to identify groupings of similar neighborhoods. The methodology assumes that these artificial statistical constructs represent individuals living in each area. In reality, of course, since the ideal of residential sorting on the basis of uniform social identity has only been partially achieved, such analysis commits aggregation errors associated with the ecological fallacy and "modifiable unit area problem" (Martin, 1991; Openshaw & Taylor, 1981; Hughes, 1991).

The marketing application of cluster models also assumes, however, that "you are what you buy" (Piirto, 1991, p. 233), or that a social identity derived from demographic data is perfectly correlated with consumption. The methodology is only useful as far as it is possible to infer the consumption habits of households in the cluster neighborhoods on the basis of consumer studies of other households who live in similar neighborhoods. Marketers, therefore, conveniently presume that society is spatially sorted by consumption characteristics, or in the words of Robbins himself that "humans group themselves into natural areas. . . . They create or choose established neighborhoods that conform to their *lifestyle of the moment*" (Robbins, quoted in Burnham, 1983, p. 92, emphasis added). This, it seems to me, is as much a normative as a positive statement.

If the marketer knows the characteristics (the "profile") and addresses of his or her best customers, and matches these with clusters, this data can be used to identify potential customers (or "prospects"), across other ZIP code areas and to increase sales (or "market penetration'') through direct marketing, media advertising, or even the location of new stores.[2] New prospects can be identified (or ''qualified'') and new customers created (or ''cloned'') in the process (Holtz, 1992, p. 311) ''with a degree of selectivity hitherto unknown'' (Robbins, quoted in Burnham, 183, p. 93).

Today, Claritas, the company which Robbins originally founded, ''crunches data on more than 100 million American households from credit bureau files, car-ownership banks, census records, purchasing surveys, and dozens of other public and private services'' (Roberts, 1992, p. 26) to produce 40 PRIZM life-style segments for every ZIP code in the United States. It is one of several competing proprietary clustering systems that perform analysis on up to 150 census variables for about 7.5 million blocks and/or 250,000 census block groups.[3] This is combined with address-matched data on individual households from various public and private sources. MicroVision (the clustering system of Equifax Marketing Decision Systems), for example, appends psychographic systems from VALS, consumer panel data from Simmons and Mediamark, and about 30 additional variables, including average installment credit, minimum bank balance, and aggregated debt burden, to produce its ''95 unique and homogeneous segments'' (Equifax Marketing Decision Systems, 1991; Curry, 1993, p. 236). MicroVision combines data from the massive Equifax Consumer Marketing Database and census data to produce clusters for every ZIP + 4 (5–15 households) in the United States.

NDL's (National Demographics & Lifestyles) Lifestyle Selector, for example, provides access to ''30 million accurate and highly deliverable names of actual consumers . . . [and] highly sophisticated customer profiling and scoring techniques designed to increase . . . response'' (Gwen Mayer, National Demographics & Lifestyles Customer Information Center representative, personal communication, September 15, 1993). It operates two clustering systems: NDL Focus, a market segmentation based on sophisticated multivariate analyses of aggregated data, and Cohort, a segmentation of individuals ''offering three-dimensional portraits of *real people.*''[4] Moreover, its Lifestyle Network database contains information on the interests and hobbies of 90 million U.S. households which allows the marketer to ''create custom mailing lists that give you virtual saturation of your prime market, while eliminating those households that do not fit your consumer profile'' (National Demographics & Lifestyles, 1993a, p. 11).

GIS

> *Revolutionary new personal computer software is now available that gives every businessperson the power to unleash the geographic information contained in every database. (Baker & Baker, 1993, p. xvi)*

The development of GIS also depended upon computer technology and related technical developments, such as relational databases and high resolution graphics. Its rapid development in the last two decades has been driven by the development of technology, and by its application in the government and commercial sectors where the primary concern has been its utility in terms of financial cost and benefit, rather than its validity, let alone its ethical implications (Martin, 1991).

There are several competing definitions of GIS, but essentially it is a combination of computerized cartography, database management, and spatial analytical tools. These three components of GIS facilitate the following three functions: the representation of the object world as proportionate symbols on a scale-transformed coordinate geography; the accumulation of relevant information about these objects in a spatial database; and the elaboration of a set of procedures for the accumulation, manipulation, and representation of this information. The graphic image remains the primary product of GIS, but unlike the conventional analog map, GIS allows the accumulation of vast amounts of geographical data and the manipulation of this data independent of the graphic image. It thus combines the power of information technology with cartography.

Although several observers have noted the lack of theory in GIS (see Lake, 1993), it involves the elaboration of a particular model of the world that is distinct from traditional cartography. The model of reality in GIS consists of discrete layers, each representing a dimension of social life, and is typically built up from a physical base, that is, the natural or built environment. The phenomena of social and natural life become "geo-objects" (Wolf, 1990), that is, objects that can be represented by points, lines, areas, and surfaces, and their identities become an accumulation of measurable attributes. The model of reality is thus literally constructed from measurable characteristics that overlap but do not interact.

Although databases and cluster analysis can be exploited by marketers independently of GIS, the capacity for spatial analysis and representation increases the power of the database, literally "turning data into competitive advantage'' (Tactics International Limited, 1991). First, in combination with center-line geographic files such as the TIGER produced by the U.S. Bureau of the Census, GDIS allows accurate location of individual addresses; second, it then provides for the aggregation of individual records identified by address into established spatial units (particularly ZIP code areas); third, it allows for the creation of new units or the hypothetical manipulation of areal units to create customized retail or consumer geographies (e.g., trade areas around stores); fourth, it provides for the reconciliation of data collected at differing scales or in overlapping geographical units (e.g., combining store trade areas with census districts, police beats, television viewing areas, ZIP codes, and postal carrier routes)[5]; fifth, it allows the exploitation of a geographical tie between disparate data sets (Moloney, 1993, p. 107) which assists in the cross-referencing of individual records and the consolidating or eliminating of duplicate records (address matching); and sixth, it provides for the visual representation of information that is both immediately meaningful and ''actionable'' to a corporate executive.

GDIS thus allows the marketer to analyze and represent information about consumers and where they live, a task that could previously be performed only by skilled technicians using mainframe computers. The first system to integrate geographic representation and analysis with consumer databases and geodemographic segmentation was Infomark Laser PC, introduced in 1985 by National Decision Systems (since acquired by Equifax). There are now a number of complete systems since the field of GDIS is increasingly integrated through licensing or reselling agreements and mergers.[6]

Tactics International Limited, for example, produces Tactician, a ''total marketing system,'' which combines in its Micro-Marketing Machine a global-capacity GIS and a million-row database that has ''snap-in'' connectivity to 45 consumer and geodemographic databases, with retail merchandising gravity models, drivetime analysis, optimized retail allocation model, and retail forecasting regression models (Tactics International Limited, 1993).[7] Conquest from Donnelly Marketing Information Services (DMIS) integrates 14 levels of census, postal, marketing, and media geographies with access to ''more than 40 industry-specific statistical and point-coded databases'' (Donnelly Marketing Information Services, 1993a) and to Donnelley's residential file of 87 million households and 140 million individuals. Compass, Claritas's desktop marketing workstation, provides industry-specific

GDIS applications for advertising, banking, cable television, insurance, and newspapers, and includes yellow page districts (YPDs) as well as the usual range of census, retail, and media geographies. Infomark is the GDIS of Equifax Marketing Decision Systems, an integrated decision-support system that combines sophisticated GIS with the MicroVision segmentation and the vast databases of Equifax.

Oasys (Opportunity Analysis System) from NDL is a market analysis system used for market segmentation, market share analysis, site evaluation, new product development, and media planning with a complete GIS. It comes with a proprietary database designed specifically for this system that draws on NDL's 30 million name consumer database and the vast data resources of R. L. Polk (including vehicle registrations). Insite.USA is the GDIS of CACI Marketing Systems, used primarily for site location and market analysis. It integrates various geographies and a basic mapping capability with the ACORN Lifestyles system, and Market*Master, a database marketing system.[8] Two other vendors offer a GDIS of more limited scope. DataMap's GDIS includes MarketPro for demographic analysis and the Datamap Locator linked to demographic data from the U.S. Bureau of the Census and the U.S. Postal Service. It does not, however, offer a segmentation scheme. Finally, the Sachs Group also offers a GDIS, including a segmentation system and the Market Planner, which incorporates GIS functions. This is designed specifically for use in the health industries, however, so I will not consider it further here.

The major GDIS presently share access to a number of commercial databases through contract arrangements; much of their own proprietary information ultimately derives from the same list brokers. Moreover, they all must share the demographic data, derived from the census and other public records, upon which the systems depend. The GIS and cluster components of their systems are unique, and the geographical and statistical algorithms are closely protected, but they perform similar functions. The differences are apparent in the know-how required, the speed of operation, and the aesthetics of the product. It is, in other words, difficult for a potential user to choose between them (Curry, 1993, p. 245; Baker & Baker, 1993, pp. 94–107). There is clearly much at stake in the attempt to differentiate these systems through product development and advertising, and a great deal of time and money is spent on promotion. While both the professional literature and the advertising that promotes GDIS in general and its proprietary systems in particular is inevitably hyperbolic, there is clearly a consistent set of metaphors which is evidence of the institutionalization of a discourse of GDIS as the "discipline" emerges from its constituent fields of marketing, social science, and technocartography (Pickles, 1991).

THE STRATEGY OF MARKETING GDIS: THE MARKETING OF *STRATEGY*

> *We make it possible to integrate demographics, specialized data, customer data, geography, mapping, and real-life marketing experience into one process called . . . "Precision Marketing." (Claritas/National Planning Data Corporation, n.d.)*

The professional literature employs many standard tropes and rhetorical devices associated with advertising in general (e.g., value for money, service, reliability, and aesthetic appearance), but it also consistently deploys a set of related metaphors that represent the functions and potential of GDIS. These metaphors are perhaps peculiarly important in GDIS because the technology is essentially a "black box" for most prospective users, in which neither the hardware nor the software are really understood (Huff, 1993). I shall identify the metaphors used to simplify and mystify the technology as the following: interiority, surveillance, position, and militarism. Each of these metaphors is itself complex, in that it is composed of a number of related metaphors, and some of these constituent metaphors are secondary concepts, in that they are derived from information science and marketing. The discourse of GDIS has, however, appropriated their meanings and associations, and they are organized into a complementary unity by the operation of the structural metaphor of *strategy,* the master trope of GDIS discourse.

A "structural metaphor" is a highly structured concept employed systematically to delineate other concepts (Lakoff & Johnston, 1980, p. 61). It need not be explicitly articulated or defined but it operates as a guide to meaning and action in the discursive context within which it operates. The aptness of such a metaphor depends upon its resonance for a community of users, and the systematic exploitation of its associations presupposes that this community shares the particular perspective on the world that their language represents.

The term "strategy" itself is used regularly in the promotional literature in both its vernacular sense, a careful plan or method to achieve a desired goal, and in its more specialist sense, the art and science of (military) command exercised to integrate and optimize the deployment of forces in the field (of battle). As we shall see, the literature is replete with tropes of gaming and militarism. These senses of "strategy," however, are not structural metaphors, but phenomenal forms of *strategy,* as the term is used by Michel de Certeau in his *The Practice of Everyday Life* (1984). *Strategy* is:

> the calculation (or manipulation) of power relationships that become possible as soon as a subject with will and power (a business, an army, a city, a scientific institution) can be isolated. It postulates a *place* that can be delimited as its *own* and serve as the base from which relations with an *exteriority* composed of targets or threats (customers or competitors, enemies, the country surrounding the city, objectives and objects of research, etc.) can be managed. As in management, every "strategic" rationalization seeks first of all to distinguish its "own" place, that is, the place of its own power and will, from an "environment." A Cartesian attitude if you wish; it is an effort to delimit one's own place in a world bewitched by the invisible power of the Other. It is a typical attitude of modern science, politics and military strategy. (pp. 35–36)

Strategy is instrumental reason, that is, a *rational logic* of human communication and action that is employed in the resolution of practical problems, and a *logical rationale* for the colonization of the lifeworld (Marcuse, 1964; Horkheimer & Adorno, 1972; Habermas, 1970). It is "technocratic consciousness" (Habermas, 1970, pp. 84–119), an ideology that represents (articulates and legitimates) the coalescence of technologies in production, consumption, and administration through which individuals are bound to networks of organizations and consumer goods, but which they cannot even imagine (Habermas, 1973, pp. 255–256). Its purpose is the reproduction of the "regime" of production—"rationalized, expansionist, centralized, and spectacular" (de Certeau, 1984, p. 31)—which imposes a system of commodities and commodified meanings upon the consumer. Under this regime, the market has become a tool for the reproduction of culture conducive to the continued expansion of the capitalist mode of production, and an autonomous civil society thoroughly colonized by the state (Marcuse, 1964). *Strategy* is a discursive field which resonates for the representatives of capital and the state, and also for those in the disciplines of the human sciences with an ideological investment in the essential governmentality of social life (Foucault 1979).[9]

The degree to which *strategy* has colonized the lifeworld is certainly debatable, and I do not share the view that consumers are dupes facilely manipulated by capital and the state (see Fiske, 1989, Morris, 1988). But, GDIS does at least promise that such strategic representation and systematic manipulation of consumers is possible. The consumer is constructed as an object that can be represented within and manipulated by the new technology. While advertisements might exaggerate the capacity of GDIS to solve problems, they presumably represent what the developer and the user would ultimately like a GDIS to be. I will examine this social construction of the consumer in the discourse of GDIS literature.

THE METAPHOR OF "INTERIORITY"

> *Architecture is no longer in architecture; the aesthetic of construction is dissimulated in the special effects of communication machines, engines of transfer and transmission. (Virilio, 1991, pp. 64–65)*

The first principle of *strategy* identified by de Certeau (1984) is the foundation of a place of power—or "interiority"—protected from the evolving circumstances of society. For de Certeau the place at the heart of *strategy* is an ordered physical space, a center isolated from the "exteriority" of everyday life-space and in a position that provides for a particular, surveillant vision of the world (historical archetypes include the castle and the panoptican).

In GDIS practice this place of power is almost perfectly realized in the concept marketed as the "War Room," the decision center developed by Tactics International Limited and applied in consultancies with several *Fortune 500* corporations (see Tetzeli, 1993, p. 91). The War Room consists of powerful computers running the proprietary Tactician software that combines the functions of Executive Information Systems (EIS), Decision Support Systems (DSS), and GIS, and "crunches" masses of geodemographic and personal data to produce maps on large video screens in presentations to corporate decision makers (see Figure 7.1).

Even the more modest desktop GDIS, however, promises an "interiority" from which to conduct strategic action. The GDIS represents a central position where the marketer can integrate data and analytical processes, literally "to internalize all your target marketing applications" (Equifax Marketing Decision Systems, 1991). Compass,

FIGURE 7.1. The War Room from Tactics International. From Baker and Baker (1993, p. 120). Copyright 1993 McGraw-Hill. Reprinted by permission.

Claritas's "targetware," for example, "provides an easy way to integrate, analyze, and map demographic, consumer demand and usage, site, and geographic data—right at your desk" (Claritas/National Planning Data Corporation, 1993). This convenient position is also isolated and secure so that corporate decision makers can plan prospecting without the competition or the consumer knowing how this is done (Garvin, 1993; Holtz, 1992; Baker & Baker, 1993). The director of business analysis for McDonald's Corporation testifies that "our desktop demographic workstation ensures the confidentiality of our marketing strategy'' (quoted in Curry, 1993, p. 268) while the promotional literature for NDL (1992b) claims that ''unlike traditional advertising programs, where the competition can see exactly what you're up to, a database marketing program is *private.* Your competition doesn't know the scope of it or what you're doing with it, and they can't respond to it until it's too late.''

GDIS is also literally represented as a construction, a ''built environment,'' consistent, of course, with the architectonic metaphor so pervasive in the discourse of information technology. It has been argued, for example, that ''what iron, steel and reinforced concrete were in the late 19th and early 20th centuries, software is now: the preeminent medium for building new and visionary structures'' (Gelerntner, 1989, p. 66). This architectonic metaphor effectively gives substance to a language, reifying the binary code that represents information as an alternate world, literally a data ''structure.'' The promotional literature of database vendors, list brokers, and GIS is replete with architectonic metaphors: databases are built and assembled, data possesses an architecture, layers are stacked, models possess an architecture, coordinate geography forms a base, and software operates on a platform.[10]

The difference with GDIS, however, is that each element or attribute in its ''architecture'' is fixed through geocoding to the two-dimensional coordinate system of the ''basemap'' or ''geographic layer.'' The abstract data structure is then anchored to a direct representation of reality, which leads to the conceit that the world of the GDIS is itself another reality. Digital Equipment Corp. (1993), for example, claims that the ''whole world'' is to be found in its GIS; Spans Map can ''discover hidden worlds'' (Tydac Technologies Corp., 1993); and CAD Images, Inc. (1993) provides a conversion algorithm for ''people who live in an AutoCad vector world.'' Infomark-GIS allows you to ''see your market through a new dimension'' (Equifax Marketing Decision Systems, 1993), and Claritas in an advertisement that depicts a vertical window frame and grill with its shadow projected onto the horizontal surface tells the reader to ''prepare to enter a new dimension'' (Claritas, 1993), the brave new world of the GDIS.

Here is the perfect edifice for *strategy,* an ironic doubling of the interiority–exteriority relationship. A representation of the ''exteriority'' of the world is interiorized in the computer. The world of the ''other'' and its identity have been captured and contained on a spatial grid by the machine technology, where it can be systematically observed and manipulated, and by the *strategy* and power on the other side of the screen. Conjuring visions of cyberspace or the "universal controller," the Cohorts database, for example, promises that "your customers . . . come to life right before your eyes" (National Demographics & Lifestyles, n.d.), while Genasys's GIS "brings data to life so you can view it, understand it, and manipulate it" (Genasys, 1993).

THE PANOPTIC METAPHOR

> *The map, a totalizing stage on which elements of diverse origin are brought together to form the tableau of a "state" of geographical knowledge, pushes away into its prehistory or posterity . . . the operations of which is the result or the necessary conditions. (de Certeau, 1984, p. 121)*

The second principle of *strategy* identified by de Certeau is that of "panoptic practice," the division of reality into spatial units for the purpose of systematic observation, the establishment of the "eye of power" (Foucault, 1980a). GDIS is a "visualization technology" (Haraway, 1992, p. 298) that possesses the power to survey, capture, and represent the world objectively from an Archimedian position outside of social reality. GDIS promises a visualization of the market terrain as a map, fixing a representation of physical and social geography, and of the identity of consumers, to a thoroughly rational Cartesian geometry. The discourse of GDIS exploits tropes of vision to promise power—from a privileged perspective the marketer attains insight, prescience, and omniscience.

It is conventionally accepted that 85% of business information has geographic attributes (Baker & Baker, 1993, p. xv), and that the map has become a critical tool for the visualization of the marketing terrain for efficient decision making. While the literature valorizes the aesthetic of "exquisitely detailed" maps produced by GDIS (Baker & Baker, 1993, p. xvii), the map is described as "almost incredible . . . wonderful" (Holtz, 1992, pp. 127–128) because of the insight it provides. With a map of the marketing "terrain," corporate executives can "see" things much more effectively than they can with traditional spreadsheets (Tetzeli, 1993), "allowing them to visualize and take advantage of options and opportunities that other data analysis techniques simply can't show" (MapInfo, 1993). GeoQuery Corporation (1992), for example, "provid[es] you with a new and much better, view," and Ramtek (1993) promises "a better view of the terrain," while NDL Focus "add[s] clarity . . . perform[s] magic and provides extraordinary insights" (National Demographics & Lifestyles, 1993b). GDIS maps provide a means "to discover 'hidden' opportunities" (Foust & Botts 1993, p. 38) or "plots" (MapInfo, 1993), and to "instantly convert mountains of data into vivid, action-oriented marketing information" (Claritas/National Planning Data Corporation, 1993). Decisions become as simple as the color coding of the maps. With UDS (Urban Decision Systems), for example,

> Now you can visualize the data you receive with a four-color Thematic Map. You can rank your neighborhoods by any data variable you choose. So you can instantly see high concentrations of income in red and low concentrations in blue. (Urban Decsision Systems, n.d.)

With this system, "targeting your market becomes as simple as 'red is hot and blue is not' " (Urban Decision Systems, 1992). The map is used to reveal the relation between geography and markets to executives, the *strategic* decision makers whose models of reality depend upon a clear legibility of functions and relations (de Certeau, 1984, p. 199). Instrumental rationality requires the operation of an *observable* cause–effect relationship as the basis for prediction and action, and GDIS, for example, provides a means "to better visualize potential markets and competitive scenarios . . . eliminat[ing] the human error (Strategic Mapping, Inc., 1993).[11] It is claimed that computerized maps can uncover otherwise obscure patterns in data, empowering the user "to spot opportunities you may have never known existed" (MapInfo, n.d.), and to explore new marketing opportunities for increased sales and profits (Baker & Baker, 1993, pp. xv, 11). With all this, it is not surprising that in the commercial sector "a map is worth a million bucks" (Stoecker, 1993, p. 231).

A familiar trope in advertising for new technologies is, of course, the futurist aesthetic. The trope is widespread in GDIS discourse, although more emphasis is placed the practical value of GDIS in equipping the user for future developments in marketing, and particularly changes in the consumer landscape, through the ability to see the future. GDIS is, for example, a means to "get the edge today for the challenges tomorrow" (Genasys, 1993). Moreover, GDIS is a tool that enables the user to actually foresee this future, for "with a clear vision of today, you can paint a clearer picture of tomorrow" (Graphic Data

System, 1993). The integrated GDIS is thus "for corporate *visionaries* who want to lead their industries into the next century" (Baker & Baker, 1993, p. 121; emphasis added).

The literature also emphasizes the importance of having the latest geographical and consumer data to support marketing decisions. Consumer data must be current, "fresh" (National Demographics & Lifestyles, 1993c), or even "live" (Thomas & Kirchner, 1991, p. 13). Thus vendors provide regular updates for their systems: Dynamap/2000 offers "real-time" maps with monthly updates of its geographical files, and Claritas's Update (instantly available through its "on-line" database) provides current estimates and projections of demographic trends, after "adjusting and confirming these estimates with primary data obtained from a network of hundreds of federal, state, and local governmental, as well as private-sector sources" (Claritas/National Planning Data Corporation, 1992). DMIS tells us what these sources may be: "DMIS continually updates its database which includes 85 million households by matching telephone book listings and automobile registrations . . . birth and death records, drivers' licenses and high school graduation records" (Donnelly Marketing Information Services, 1992, n.p.). De Certeau argues that *strategy* seeks to transcend or master temporality by its occupation of a place and organization of space around a stable position conceptually independent from the contingencies of exterior circumstance. This fixed position is a place from which to survey the territory and to safely predict changes in the exteriority, or "to run ahead of time by reading a space" (1984, p. 36).

GDIS also promises a "god's eye view" of the world (Bondi & Donosh, 1992, p. 202), a privileged, multiple-perspective vision of the world of consumption. With Spans Map, for example, "four simultaneous views can be interactively explored to uncover hidden facts or relationships" (Tydac Technologies Corp., 1993), while Tactician allows you to "move between countries . . . [and] zoom in close enough to see a shopping mall and zoom out again to see the world, in a response time of less than a second" (Tactics International Limited, 1991). The user has the whole landscape and population under surveillance, no longer by searching out over the castle ramparts or peering through the grill of the prison bars, but in the panoramic convenience of geocoded databases and GIS. GDIS is thus the perfect panoptican—or perhaps Superpanoptican "a system of surveillance without walls, windows, towers, or guards" (Poster, 1990, p. 121).

Observation and analysis of individual behavior do not depend upon the confinement of the observed within an institutional structure, nor the potential presence of an observer as witness of behavior to maintain discipline. In fact, the individuals who complete survey forms about their life-style and product preferences, or who automatically provide information as they log on to electronic services, effect a continuous self-monitoring while providing information to the remote analysts. Deleuze (1992) thus identifies the eclipse of Foucault's "disciplinary society," in which social discipline is based on enclosed institutions or "molds" (school, prison, factory, family, etc.), by the *society of control,* in which social control is exercised via "modulation," the universal application of which is the computer which tracks and is capable of locating the individual at any given moment/position.

In marketing discourse, the consumer is, of course, stereotypically female (Ozanne & Stern, 1993) and several of its tropes—from seduction to market penetration—are overtly (male-hetero) sexual. The combination of the discourse of marketing with a visualization technology makes GDIS, not surprisingly, a perfect example of what Donna Haraway (1992, p. 295) calls "technopornography." The technical apparatus is eroticized in an aesthetics of power, speed, and control. The GIS of GeoQuery, for example, provides a "snooper tool," a voyeuristic facility that displays "everything from sales figures to phone numbers" (GeoQuery, n.d.); Compass is a "powerful tool" that "will improve your actual performance" (Claritas/National Planning Data Corporation, 1993); Tactician has "extraordinary speed and huge capacity" (Tactics International Limited, n.d.); and Conquest—well, the name says it all. In addition, as discussed below, GDIS also promises to provide "intimate portraits" of consumers and to "uncover their desires" to the marketer.

Unlike conventional pornography (but not unlike the emerging field of "on-line" pornography), the "technopornography" of GDIS is interactive. GDIS allows the user to "query the data," to change the projection, perspective, or position on the map, instantaneously altering its geography. Sitting in front of the screen the (imputed) male marketer can, for example, play with the world within at whim, and with "the touch of a single button . . . insert data into each site in the trade area model" (Tactics International Limited, 1993). GDIS lends itself to instrumental perception of the world through its "what-if" and "show-me-where" scenarios (Strategic Mapping, Inc., 1993) in order to redesignate trade areas or otherwise alter the retail geography of reality. Like a video game, the GIS component of GDIS can "both simulate spatial environments and impart complete control to an operator'' (Dobson, 1993, p. 20).

The manipulation of representational space is made possible by the reassignment of boundaries which presupposes, of course, a prior spatialization, the imposition of order realized through division of the world into domains for the purpose of effective administration. The manipulations are achieved by stretching existing boundaries and arbitrarily reapportioning characteristics based on the assumptions of uniform distributions of characteristics within the established areas. The practice depends upon the orderly geography established by, for example, the U.S. Bureau of the Census (census tracts and blocks), the U.S. Postal Service (ZIP code areas), state and local governments (electoral districts), social service administrations (police beats and health districts), and various market research corporations—for example, the television viewing areas of Arbitron (Areas of Dominant Influence) and Nielson (Designated Market Areas).

In Lefebvre's (1991) terminology, these geographies are ''representations of space,'' spaces that are conceived entirely for the purpose of social control; and they are essentially independent of everyday life and the ''representational spaces'' in which people invest meaning. The spaces of GDIS are administrative grids, systems of boundaries that define functional regions within the geopolitics of power (see Foucault, 1980b, p. 77). Moreover, they constitute a normative model for the representational space of an instrumental rationality, a model for the world as conceived and constructed by technocrats and design professionals for the purpose of efficient administration by bureaucratic and corporate power. In this world, social life is spatially organized and efficiently segregated so that difference only occurs at the boundaries that separate social spaces, and social identity is constituted in and perfectly predicted by residence. If social life can be divided into homogeneous areas by discrete boundaries, it becomes possible to identify and describe people by place, and vice versa, an ability that obviously increases governmentality by both capital and the state. That is, geographical locations can be described in terms of the characteristics of residents and particular types of people can be easily located. Also, as marketers speak of the purchasing power and consumption habits of neighborhoods rather than their inhabitants—thus, for example, a ''neighborhood eats croissants'' (Curry, 1993)—identity is subsumed by territory and social space is reified.

THE METAPHOR OF POSITION

> *With Conquest you can analyze and display your target audience by number, penetration, index or however you want. (Donnelley Marketing Information Services, 1993a)*

The third principle is the attribution of each element of social life to its appropriate position within the "exteriority" for the purpose of management. This process depends upon "the power of knowledge" over objects and territory (de Certeau, 1984, p. 36), the constant monitoring of behavior with the testing of predictive models, and the assignment of each element and activity of social life to its proper place (the "Cartesian attitude"). GDIS "assigns" position to consumers in both senses of the word, locating them simultaneously in social and geographical space.

The first moment of position is the systematic observation of the "other," that is, social subjects and their lives. Consistent with the contemporary "cult of information" (Roszak, 1986) or "mythinformation" (Winner, 1986, p. 105)—DMIS has even registered the term the "Age of Information" (Donnelly Marketing Information Services, 1993b)—GDIS discourse commits to a belief that marketing power is simply proportional to the quantity and quality of information available. It is not surprising, since GDIS sells the ability to accumulate and manipulate information about consumers, that GDIS advertising should argue that "your business decisions are only as good as the information you use to make them" (National Decision Systems, 1993). Even the more academic GIS literature argues that information is an inherently valuable commodity, particularly information about people that has a geographical or spatial component (Rhind, 1992a, p. 263). Geographical market information is touted as "the most important tool of all . . . in today's competitive business climate" (Donnelley Marketing Information Services, 1992).

Multivariate statistical analyses can reduce any amount of data to "significant" components, so it is not surprising that the promotional literature privileges quantity. There are numerous references to the staggering numbers of records in MCFI systems and the huge capacity of GDIS for storing and manipulating this data in "million row" GDIS databases (Tactics International Limited, 1993). The goal is to accumulate as much information as possible, for "with complete empirical knowledge of people's retail activity, we can better predict consumer behavior'' (quoted in Thomas & Kirchner, 1991, p. 73). Andrew Garvin, president of FIND/SVP, citing Bacon's old adage, recognizes that knowledge means power, and has recently published a guidebook to raise ''information consciousness'' and to help the executive become ''information confident'' (Garvin, 1993).

Complete empirical knowledge of the consumer is attained by combining vast amounts of data from diverse sources which systematically observe the characteristics of the population and monitor the behavior of customers. Part of this data has been collected and used by the administrative apparatus and its sciences, that is, the data continuously or regularly collected by government and related agencies for their various purposes. It has only been since the development of GDIS, however, that the vast amounts of information contained in these databases, which are generally protected by confidentiality rules, have been used by marketers. This information is combined with proprietary data collected by the multi-million-dollar information industry that keeps track of consumers and manages massive databases on the behavior of various categories of consumer.[12] The ''complete consumer'' (Francese, 1988, p. 11) generates masses of personal information by subscribing to media services, responding to surveys, and making regular purchases (using credit cards and filling in warranty cards), leaving a trail of paper and electronic data, all of which can be effectively logged, cross-referenced, and analyzed for use by marketers.[13] GDIS companies sell data from both of these sources for use on their systems in various forms, including published reports, diskettes, CDs, tapes, or more recently on-line, so that they can be conveniently loaded directly into the GDIS for analysis. The volume and detail of data available is extraordinary.

The second moment of position is the analysis of social life and the prediction of behavior based on relationships between variables in this data. Sophisticated statistical analysis combines aggregated socioeconomic data on neighborhoods with individual data on customers to produce models of consumer behavior, or ''consumption classes.'' Through ''intelligent'' address matching each individual or household is precisely located on the map, which then becomes the basis for inferences about the characteristics of other individuals or households. All residents in all neighborhoods in the United States can then be assigned to a particular consumption class, and this map of the geography of consumption becomes the basis for the prediction of subsequent consumer behavior and management of marketing campaigns. Whether these models of sociospatial positioning are accurate or not, this surely promises to be the ultimate instrument of social control.

Ideally, of course, the marketer would be able to obtain personal information on all consumers and to reach them directly (hence the promise of interactive television), but the information industry itself recognizes that societies will inevitably become more protective of the confidentiality of personal information as the population becomes more aware of the extent of its use and abuse in marketing (Rhind, 1992a, p. 263; Hughes, 1991, pp. 431–442). Database managers have adopted various strategies to increase voluntary participation (Goss, 1993), but they must ultimately rely on geographical inference to identify "prospects" despite its problems. The underlying assumption is that "we know who you are" because "we know where you live."

Thus neighborhoods are classified according to their general consumption characteristics inferred from an analysis of aggregated data. The neighborhoods are given appropriate stereotypical names, typically indicating their geographical location and ranking according to their general levels of consumption. PRIZM, for example, includes "Furs and Station Wagons" and "Pools and Patios" in its suburban segments and "Shotguns and Pickups" and "Hard Scrabble" in its rural groups. ACORN includes "Trend-setting suburban neighborhoods" (B Group) and "Middle Americans in new homes," and PRIZM includes "Hispanic and Multiracial Neighborhoods" and "Families in Pre-War Rentals." MicroVision organizes its 50 clusters into 9 primary groups for which it provides portraits of its typical representatives in front of the homes in which they are stereotypically housed (see Figure 7.2).

Although GDIS positions all consumers, and predicts behavior at all "levels" of consumption, it is clearly most interested in locating those who most conform with the normative goals of behavior prescribed by marketers, and ultimately with the regime of production. The ideal consumer in this sense is the life-style-conscious individual, who, like the marketer, conceives of identity in terms of consumption characteristics. Typically also, renters of lists that contain big-spending and self-conscious life-style categories must pay a premium for these names and the promotional literature inevitably emphasizes GDIS's capacity to identify and target affluent consumers.[14] Claritas's ZIP Quality Index, for example, is "a weighted composite of education and affluence variables that permits clusters to be ranked and grouped" (Curry, 1993, p. 239). The ability to discriminate between individuals in terms of the degree to which they act in accord-

MICROVISION GROUPS
The 50 market segments of MicroVision are organized into one of nine groups*. Since each group is comprised of segments that have similar characteristics or habits, you can easily target several market segments that share common lifestyles and purchasing behaviors.
ACCUMULATED WEALTH (MVG1)
SEG. # NAME
1 Upper Crust
2 Lap of Luxury
3 Established Wealth
4 Mid-Life Success
5 Prosperous Ethnic Mix
6 Good Family Life
14 Middle Years
MAINSTREAM FAMILIES (M
SEG. # NAME
10 Home Sweet Home
11 Family Ties
16 Country Home Fami
17 Stars and Stripes
18 White Picket Fence
22 Aging America
23 Settled In
35 Late-Life Laborers
38 Back Country
YOUNG ACCUMULATORS (MVG3)
SEG. # NAME
9 Building a Home Life
19 Young and Carefree
25 Bedrock America
28 Building a Family
MAINSTREAM SINGLES (M
SEG. # NAME
8 Movers and Shaker
12 A Good Step Forwar
15 Great Beginnings
32 Metro Singles
34 Books and New Rec
39 On Their Own
40 Trying Metro Times
CONSERVATIVE SENIORS (MVG
SEG. # NAME
7 Comfortable Times
20 Social Security
21 Sunset Years
30 Retirement Age
31 Golden Times
ASSET-BUILDING FAMILIES (MVG5)
SEG. # NAME
27 Middle of the Road
29 Establishing Roots
SUSTAINING ETHNIC FAMILIES
(MVG8)
SEG. # NAME
24 Metro Minority Familie
41 South of the Border
42 Hanging On
43 Low-Income Blues
44 Hard Years
46 Difficult Times
CAUTIOUS YOUNG COUPLES
(MVG7)
SEG. # NAME
26 The Mature Years
33 Living Off The Land
SUSTAINING SINGLES (MV
SEG. # NAME
13 Successful Singles
36 Metro Ethnic Mix
37 Moving Ahead Minor
45 Struggling Minority
47 University U.S.A.
48 Innercity Singles
*Segments 49 and 50 were not placed in a group due to their non-homogeneous nature.

FIGURE 7.2. The MircroVision clustering system, the nation's most widely used customer segmentation and targeting system, developed by Equifax National Decision Systems. From Equifax Marketing Decision Systems (1991). Copyright 1991 Equifax Marketing Decision Systems. Reprinted by permission.

ance with the wishes of those with power has been noted to be an important function of information technology (Kling, 1985, p. 271).

Despite the distancing and detachment that is essential to GDIS *strategy,* it nevertheless promises intimate knowledge of consumers through continuous and detailed observation of social life. GDIS is essential to the new "value," "handshake," or "relationship" marketing (National Demographics & Lifestyles, 1993d) in which marketers identify "prospects" and appeal to their individual characteristics, and subsequently maintain detailed information on purchasing habits, product satisfaction, and changes in household composition, life-style, or other characteristics that might lead to the sale of further products or services (Holtz, 1992; Hughes, 1991). The goal is to monitor consumers and to obtain sufficient information on their everyday lives to be able to anticipate their consumer needs and desires. GDIS literature thus promises to "restor[e] the personal contact with customers that we all enjoyed in the early years of America" (Hughes, 1991, p. 516) and to reproduce the intimate relationship that existed before the mass-marketing era when seller and buyer liked each other, and the traditional corner grocers personally knew and cared for the lives of all their customers (Holtz, 1992). Database marketing will allow marketers to "get acquainted" with their customers (Holtz, 1992, p. 3) and even "mak[e] America a better place to live!" (Hughes, 1991, p. 24). DMIS's Conquest, for example, promises that "you will communicate with your customers more effectively and efficiently because you'll be speaking their language" (Donnelley Marketing Information Services, 1993a) and NDL's Focus lets you understand your customers better, so you know who they are, what they do, how they spend their time and money. Focus will even give you "the most comprehensive 'living' portrait of your customers. [It] takes an important step beyond demography and geography. In remarkable detail, it reveals their lifestyle and mindset" (National Demographics & Lifestyles, 1993b).

In each of the clustering schemes the demographic and consumption characteristics of each cluster are described in detail, including incomes, family status, expenditures, and media habits. Other characteristics such as political orientation and even personality are often inferred from this data. The inference is taken to the extreme in NDL's latest cluster system called Cohorts, in which each segment is given a name or names according to the first names of individuals in its demographic and consumer databases that are most typically associated with each cluster. A brochure promises that with this system "you meet your customers on a first name basis'' and the marketer is provided a ''vignette'' as an introduction to each of the 33 ''characters''—individuals or couples—which is a soap opera-type scene of an intimate moment in their consumer life (see Figure 7.3). It is claimed that ''as you get to know [the cohorts], you'll discover these are all people you know: your friends, neighbors, relatives, and—most importantly—your customers'' (National Demographics & Lifestyles, n.d.).

The rhetoric of intimacy is extremely ironic given that GDIS reduces identity to a set of measurable characteristics in order to predict behavior, and is only interested in individuals as ''target addresses'' or ''prospects'' that bear these characteristics and potentially the desired market behavior. It is not people with ''real'' names that marketers get to know, but statistical constructs derived through correlations of socio-economic and demographic variables aggregated at levels of census geography, but made meaningful to the marketer by ''clever'' or ''catchy'' names (Hughes, 1991; Holtz, 1992). The ''intimate portraits'' that the GDIS systems paint (National Demographics & Lifestyles, 1992a) are not of actual individuals but of stereotypes that appeal to marketing executives—one could hardly imagine a method to produce anything less intimate with or representative of the real everyday lives of consumers.

This is obviously ideological and one might be justifiably skeptical whether a cluster analysis of aggregate data on neighborhood socio-economic and demographic characteristics with limited information on life-styles, typically obtained from warranty cards and follow-up surveys, could be so revealing of the personalities and everyday lives of consumers. However, it seems far less likely that GDIS vendors are consciously fooling the hapless marketing industry than that they share its view of consumer identity as an aggregate of measurable social characteristics defined by place of residence, a view consistent with their intent to represent and control social life for instrumental purposes. Construction of a subject–object dichotomy through the establishment of position is an essential element of *strategy* (de Certeau, 1984, p. xix).

GDIS does not conceive of the individual as a self-constituting agent as much as a constellation of statistics that is produced by the regime of production. GDIS, like positivist social science and planning, is a technical-calculative practice (Feenberg, 1991, p. 184) that decontextualizes and reduces individuals to the measurable characteristics that they bear (see also Lake, 1993), such that they exist only in so

Margot & Elliott

Married Households
Homeowners
Median Age = 56
Median Income = $73,366
% U.S. Households = 8.1

The second largest group of the population, this Cohort is comprised of **established, affluent, empty-nest** households. Over 40% of all U.S. households reporting an income **over $75,000** are found in this Cohort. The vast majority are still in the work force (almost exclusively in **white-collar occupations**), and over half are **dual income** households. Their lifestyles have a distinctly **urban** character (over 80% of these households appear in large metropolitan areas). They are not averse to using credit to support their sophisticated interests, which include **foreign travel, wine connoisseurship, cultural activities,** and **gourmet cooking.** They are active investors in **stocks, bonds,** and **real estate.** While they are not especially active in the great outdoors, they do pursue **physical fitness.** Their sporting activities tend toward club sports like **golf** and **tennis,** and over half of them **walk for their health,** making them the top Cohort for that activity. **Avid book readers,** they also find time for **TV sports** and visiting with their **grandchildren.** They are the leading Cohort for both foreign and domestic **luxury car** ownership. They are most over-represented on the **west coast** and in the **northeast.**

Cohort 1

Vignette

Margot checks the clasp on her heavy gold necklace one last time and smoothes it so it lies flat. This is their third fund raiser this month, but for the first time she's ready before Elliott, thanks to his late tee-off. She glances at the clock. He'd better get out of that shower soon. It won't look good if the chairman of the board is late for the symphony benefit.

She reads the day's mail while she waits. A letter from their daughter, half crazed with midterms. Margot chuckles at her rantings—you'd think she was the first person to endure the rigors of med school. She puts the letter aside. Credit card bills, a catalog, a postcard from their son on winter break in Costa Rica. She checks the postmark. Only took three months to get here.

Elliott walks into the room, taking a final tug on a perfect bow tie. "Have you seen my cummerbund? Wow, you look great, hon." He bends to kiss her cheek.

Margot smiles. "The red one's on the bed, the black one's on the chair. Take your pick."

"Thanks."

Smoothing his wet hair, Elliott heads back to the bedroom. Margot picks up a magazine. In a minute he'll be back for help with his cuff links.

***FIGURE** 7.3.* A "vignette" of NDL's "established, affluent" Cohort. From National Demographics & Lifestyles (n.d.). Copyright by National Demographics & Lifestyles. Reprinted by permission. NDL provides comprehensive database development, database consulting, research, and direct marketing services to leading consumer goods companies. NDL's targeted list performance services improve direct mail response.

far as they can be represented and controlled as social objects by the knowledgeable subject (the market researcher). GDIS technology empowers the marketer to isolate components of an ascribed identity and reconstitute them into ideal types, and it does not matter much whether these reconstituted identities are "accurate," nor whether they help us understand consumer identity, but only that they are able to predict behavior sufficiently well to be profitable. The margins of error are generous: Given that responses to unsolicited mailings are about 2%, even a small "lift" in responses due to a segmentation of the market will increase returns (Hughes, 1991, pp. 245–248).

Several other observers have seen in the rapid institutionalization of information technology in general, and GIS in particular, both in the private and public sector, a "resurgence of positivism" (Pickles, 1991; Lake, 1993), and a renewal in the "faith" of objective social science, committed to the quantitative modeling of behavior and the rational management of social life. GDIS is an exemplar of this faith in *strategy:* Its method is statistical, and its power lies in dividing reality into units that conform with industrial or administrative production, a reality that consists only of the material and measurable world (de Certeau, 1984, pp. 34–35). It represents this reality in order to manipulate it, and its ultimate goal is to impose upon consumers—to "sell" to them—the preferred ways of using and making meaning from products consistent with the interests of the dominant order of production (de Certeau, 1984, p. xii).

"EXTERIORITY" AS THREAT: THE MILITARY METAPHOR

The final metaphor, that of militarism, is not a principle of *strategy,* but its antecedent, for historically the discourse of the administrative sciences emerges from that of militarism—in the words of de Certeau, "the model was military *before* it became scientific" (1984, p. 38; emphasis added). The discourses of science and of the military continue to overlap and cross-fertilize, however, for they remain essential elements to the master discourse of *strategy.* The literature of the recently developed administrative science of GDIS, in particular, is replete with military metaphor. Information technology and computer-assisted cartography were developed partly under generous assistance from the military which saw in them potential applications for military intelligence (Roszak, 1986). Indeed, one of the most publicized "successes" of the contemporary information and surveillance technologies has been their application in decision support in the Persian Gulf War (for a critique, see Smith, 1992), and several recent advertisements for GIS explicitly and implicitly capitalize on this role.[15] The trope of marketing as war is also well established, as evidenced in the notions of "targeting" customers and "battling" for customer loyalty. The prevalence of military metaphors in GDIS discourse is partly due to its antecedents in these fields.

First, there is the obvious use of militaristic metaphors in naming the components of GDIS and describing its functions and attributes. For example, GDIS companies have taken names such as Tactics International or Strategic Mapping, while GDIS software is registered under names such as Conquest, Tactician, and the War Room. Consistent with this convention, the instruments of GDIS become sophisticated weapons and software becomes "targetware" for use in the battle for customers: GIS is described as "the strategic business weapon" by *GIS World* (1992) and Atlas-GIS is advertised as a "secret weapon" that allows the very serious marketer to target the innocent child (see Figure 7.4). Similarly, desktop marketing systems are said to possess a "quick-strike" capability (Thomas & Kirchner, 1991, Preface), consumer information is "captured" in databases, customer lists are "targetable," users can "zero in" to "hit targets" (GIS Research Corporation, 1993), and marketers fight "data wars" (Rhind, 1992b; Tosta, 1993).

Second, and consistent with the trope of GDIS as a weapon, there is the metaphor of precision, a characteristic of information technology also highly valorized—and perhaps equally fraudulently—in recent military conflicts. GDIS in general allows a "bullet rather than shotgun approach to marketing" (Curry, 1993, p. 202), and Claritas claims that Compass "make[s] it possible to integrate demographics, specialized data, customer data, geography, mapping, and real-life marketing experience into one process called . . . 'Precision Marketing' " (Claritas/National Planning Data Corporation, 1993), a concept it has registered. The capability for precision is used, as in military conflict, to locate targets in geographic space, in this case customers and their street addresses, with "pinpoint" accuracy (see Figure 7.5). Gary Hill, president of Claritas, claims that "if you're trying to find a person with particular attributes, we can point you to his doorbell" (quoted in Roberts, 1992, p. 26), while Pinpont Analysis Ltd. is creating a 1m grid for every property in Britain that receives mail—"the most accurate

Secret Weapon

Gaining a competitive advantage has never been more difficult. Intensified competition, increasing emphasis on cost control, and shifting market demographics are redefining traditional sales and marketing challenges. Circuit City Stores meets the challenge with a powerful secret weapon. . Atlas Software from Strategic Mapping, Inc. Accelerating their nationwide expansion in consumer retailing, Circuit City is solving complex marketing challenges for site selection and targeted marketing using Atlas Software's desktop mapping solutions.

"Atlas Software allows us to better visualize potential markets and competitive scenarios. It eliminates the human error of manual methods. Now we can analyze and select prime store locations in a fraction of the time." -Julie Scott

Rated #1 Mapping Software by the Editors of INFO WORLD 12/7/92

Atlas Software is the overwhelming choice of marketing professionals like Julie Scott who need to maximize location-critical marketing decisions. Atlas Software is setting a new benchmark for power and speed in a wide-range of applications such as **Target Marketing, Territory Analysis, Distribution Planning, and Site Selection**...and making the critical difference for over 40,000 users worldwide.

Atlas Software runs on your existing PCs and Macs. It integrates digital maps with common database information, importing data formats such as ASCII, .dbf, Excel, and Lotus 1-2-3, as well as popular geographic file formats. It swiftly executes geographic "what-if" and "show-me-where" analyses right on your desktop. And it produces presentation-quality maps and reports that visually support location-critical decisions.

If you need precision sales and marketing analysis, get the industry's leading desktop mapping solution. It's fast, easy, and cost effective. And it's rated #1 by the most respected names in the business. Call us today and we'll send you a free demo package for Atlas Software... it won't be a *secret weapon* for long.

800-472-6277

Strategic Mapping, Inc.
3135 Kifer Road
Santa Clara, CA 95051
TEL: 408-970-9600
FAX: 408-970-9999

FIGURE 7.4. GDIS as a "secret weapon" with its operator and target. From Strategic Mapping, Inc. (1993, p. 4). Copyright 1993 Strategic Mapping, Inc. Reprinted by permission.

Their idea of precise geocoding.

Matchmaker/2000® for Windows™ is the only geocoding system that matches street address ranges to latitude and longitude coordinates. So you end up with a more precise and useful picture of where your customers and prospects are located. Matchmaker/2000 offers nationwide street coverage, with more than 12 million address ranges. Other programs offer only half as many. Matchmaker/2000 is continuously updated. So your data is always current. And you'll achieve the highest match percentage available in the industry today. Matchmaker/2000 is an invaluable tool for market penetration studies. Point and cluster evaluations. Sales effectiveness analyses. Scheduling and routing. And custom zone creation. You'll work smarter. And faster. The program is offered with a range of expandable and upgradable database options to meet your specific budget and application. Contact Geographic Data Technology, Inc., 13 Dartmouth College Highway, Lyme, NH 03768-9713. Or call 1-800-331-7881, x1101.

GEOGRAPHIC DATA TECHNOLOGY, INC. **1-800-331-7881 x1101**

FIGURE 7.5. The promise of pinpoint accuracy for marketing applications. From Geographic Data Technology, Inc. (1993, p. 56). Copyright 1993 Geographic Data Technology, Inc. Reprinted by permission.

system for locating every address in the country'' (Catrell, Dunn, & Boyle, 1991). Tactician, it is claimed, ''maps from the world to the street corner and smaller'' and its ''street level precision'' provides the ability to ''zoom in close to see a shopping mall and zoom out again to see the world, in a response time of less than a second'' (Tactics International Ltd., 1991). In such a display, each element of the world, each consumer address, is a target sitting on precise coordinates and waiting to be targeted by the marketer.

Third, there is the metaphor of military intelligence and reconnaissance. The trope of the intelligent machine derives from information technology, although ''intuitive models,'' ''functionality,'' ''intelligent GIS,'' and ''smart mapping'' also echo some of the buzzwords used to describe military technologies in recent geopolitical conflict. The origins of the trope of reconnaissance are harder to place, but seem to derive from the notion of the marketplace as a hostile terrain or battlefield. Thus Compass by Claritas ''leads you through the marketing information wilderness'' and ''successful marketers must know the territory'' (Claritas/National Planning Data Corporation, 1992), while Scout by NDL is a ''marketing intelligence service . . . that lets you know the territory'' and helps ''keep track of the consumer'' (National Demographics & Lifestyles, 1992b). This trope is also, however, consistent with the metaphor of colonization or military occupation which occurs elsewhere in the literature. GDIS is described as a force gradually colonizing the business world by moving from a frontier to a lonely outpost (Francica, 1992), and software packages have been registered with names such as Conquest, GeoQuest, and DataQuest (a trend, of course, also echoing the numerous computer games involving intelligent and strong warriors pitted against dragons and other forces of evil). The notion of the market as a terrain to be settled is a recurring theme: ''The once familiar landscape for marketers of durable goods has all but vanished. What seemed a vast and expanding frontier of consumers eager to purchase new products has suddenly flattened out'' (National Demographics & Lifestyles, 1992). The terrain and object of conflict, however, has seemingly shifted from the marketplace where marketers battle each other for the loyalty of customers to the consumer's mind (Ries & Trout, 1986). Marketers are now more overtly conscious of the need to create meaning and identity, or literally to colonize the lifeworld (Habermas, 1970).

Finally, and consistent with the notion of colonization, there is the trope of mining or the exploitation of the resources of ''the territory.'' The search for new customers through GDIS is called ''prospecting'' (Holtz, 1992; Hughes, 1991), and GDIS is described as a ''divining rod that will lead to a reservoir of untapped potential'' (Reisberg & Gilbert, 1991, p. 59). Similarly, although unfortunately the metaphors are mixed, Claritas, ''the nation's largest consumer and business data mill . . . , sits on a gold mine of information'' (Roberts, 1992, p. 26), while the maps and charts produced by NDL's Scout software ''show marketers where gold is today'' (National Demographics & Lifestyles, 1992b). Military occupation of the territory evidently yields up its wealth.

CONCLUSION

> *The reified models of the sciences migrate into the sociocultural life world and gain objective power over the latter's self-understanding. (Habermas, 1970, p. 113)*

The discourse of GDIS is an apotheosis of instrumental reason. It promises to measure, represent, and classify consumer identity with the explicit intent to predict and manipulate behavior. It reduces the complex multidimensional character of human identity to a set of correlated variables which are then aggregated into abstract constructs or clusters based on place of residence; these clusters are then reified into ideal types of ''consumer neighborhoods'' used to predict consumer behavior, which is in turn reduced to a simple scale representing the likelihood of purchasing a particular commodity.

Again, one might doubt the accuracy of this model and dismiss the claims made in the literature as hype. There are several objections to such an argument. First, although there is undoubtedly some self-promotional exaggeration of the capabilities of GDIS, the testimony by professionals on its utility in marketing needs to be considered—the systems are apparently working sufficiently well for *them* (for case studies of its application, see Thomas & Kirchner, 1992) and its promise is sufficient to cause a revolution in marketing (Baker & Baker, 1993, p. xvi). Second, if the application of GDIS continues to expand, one wonders what effect it might have on the retail landscape, on the nature of consumption, and on the constitution of consumer identity. I have speculated elsewhere that GDIS may be part of a self-fulfilling prophecy (Goss, 1993), since marketers define social identity in terms of unique constellations of consumer goods consistent with idealized life-styles and then attempt to sell back these identities to consumers in packages of complementary consumer goods. In this conception, identity is also, however, defined by geographical location, since it is presumed that similar people live in similar places (this is a rational ideal shared by *all* administrative sciences). Just as the logic of planning ultimately may make reality out of its assumptions, so marketers too, through their monopoly of market information and control over the means of production of social identity, can perhaps engineer the regularity they so desire. That is, by selling the means of production for *only particular* identities in *particular* places, they may create place-specific consumer identities. As was Foucault in his work on the disciplinary society, I am concerned, therefore, not so much with the invasion of privacy perpetrated through this new surveillance technology, but with the manner in which systematic surveillance and its discursive regime ultimately constitutes social subjects (see Foucault, 1986, p. 5).

This discourse of *strategy* is itself strategic, in that it offers a vision of a rationalist world that is ultimately also a means toward realizing that end. In the meantime, however, it reveals its own inadequacies, for it recognizes in its taking up of position outside of the exteriority of the lifeworld, in its objectification of the consumer, and above all in its representation of the consumer's world as hostile to its own power, the very fact that however rationalized and strategic marketing has become, the act of consumption still retains a contingent character that *strategy* does not control. The regime of production, both as corporate capital and rational government, seeks to impose order upon consumption, that is, a geostrategic technical control over the everyday life of the consumer, through the collection, consolidation, and circulation of information necessary to predict behavior. It attempts to pin down consumer identity with elaborate models and the very latest statistical data, and ultimately to impose a system of commodities upon consumers, and with these goods the means by which they can construct the appropriate identities for themselves.

Consumption, however, is essentially performative, in that its meaning and purpose is created only in the act and is never quite predictable or controllable. Although consumers have nothing but the commodified means of production of identity available to them, they select from this material in heterogeneous ways, appropriating various elements from the imposed product world in order to create and communicate their own everyday lives and identities. Consumer identity is always, therefore, becoming, always emergent (Slater, 1992, p. 190), and consumption is also always, potentially at least, an "act of inversion'' (de Certeau, 1984, p. 13). Although it cannot threaten the stability of the regime of production, consumption subverts its rationality and authority by playing with matters it takes deadly seriously, by taking the material that marketers provide in earnest and using it, often playing with it, to its own ends. Thus the everyday ''escapes'' (Blanchot, quoted in Morris, 1992) from *strategy.*

The response of strategy to the tactics of everyday life (de Certeau, 1980) is to fix the meaning of consumption and locate the consumer objectively in social and geographical space. GDIS as part of this *strategy* standardizes consumer identity by the place in which consumers live and the products they use, and uses this information to further predict the uses that consumers will make of commodities. GDIS with its origins in the discourses of information technology, automated mapping, marketing, and the social sciences, represents a unique convergence of the discourses of *strategy* which focus on the control of the individual and constitute the consumer as a threat to its rationality. In the discourse of GDIS the consumer is therefore constantly constructed as an exterior object to be captured, studied, reduced, and targeted by the operator, in other words, as the *enemy* of the intelligent machine.

NOTES

1. This was the catchy title of the exhibit of Claritas's COMPASS GDIS display in the ''Information Age'' exhibit at the Smithsonian Institution's National Museum of American History.

2. Retailers have various means of researching their customers from point-of-purchase records to warranty ''tip-ins.'' The latter are, in fact, likely to be managed by list brokers (see Goss, 1993). List brokers claim that they will already have detailed information on 15–20% of a retailer's customers.

3. CBGs are the smallest units for which relatively complete statistics are available since the census suppresses variables at block level to maintain confidentiality. There are about 300 households per block group.

4. NDL's *ZIP Screens* are analyses of ZIP code areas used to identify prospects. The analysis includes log linear modeling, logistics and multiple regression, Chi-squared automatic interaction detection, and discriminant analysis (National Demographics & Lifestyles, 1991).

5. This, however, introduces another problem specific to spatial analysis. It is not possible to reassign aggregated values unless they are distributed homogeneously across the area (Flowerdew & Openshaw, 1987, cited in Martin, 1991, p. 142).

6. For example, Claritas purchased National Planning Data Corporation

(NPDC) and discontinued its GDIS (PrimeLocationX) in 1991, and Strategic Mapping Inc. has recently reached agreement with A. C. Nielson to purchase DMIS from the Dun & Bradstreet Corporation (Geoforum, 1993).

7. "Distance, competition, price, quality, comfort, convenience and range, as well as demographics, expenditures and clusters—all come together to determine how profitable your business will be" (Tactics International Ltd., 1993).

8. CACI has said, however, that it is phasing out Insite. USA because more sophisticated GISs have been developed by other vendors, and it will now be a reseller of Strategic Mapping products (sales representative, personal communication, December 3, 1993),

9. This chapter focuses on the commercial applications of GIS, but its widespread use in the public sector should also be cause for concern.

10. According to the GIS promotional literature, for example, Aerial Data Reduction Associates, Inc. (ADR) "build[s] Geographic Information Systems"; Camp Dresser & McKee, Inc. (CDM) provides Geographic Information Architecture; GeoVision "constructs an advanced data architecture"; ST Systems Corp. (STX) vends Geostack software; and Tydac Technology Corp's InFoCAD (1993) "provides the user with a *platform* to *build* custom applications . . . [and its] data *structure* utilizes an '*Open Architecture*' allowing a seamless integration of different data *models* on which to *build* database applications."

11. Deetz (1990, p. 55) argues, rather ironically, that intuitive decision making in business is more usual than "rational" decisions based on codified information and models of reality. Nevertheless, this is clearly the privileged form so corporate decision makers will often use their models as a means of post hoc "rationalization."

12. For example, a single company sells data on segments that include, among many others, "The Youth Market," "The New Baby Boom," "The Maturity Market," "The Over-45 Women's Market," "Hispanic Consumers," and "Black Consumers and the Catalog Industry" (FIND/SVP, 1993, pp. 4–5).

13. Much of the information on life-styles (or, more accurately, consumption styles) in MCFI is derived from warranty cards and follow-up surveys administered by database managers and list brokers on behalf of retailers. The questionnaires typically contain both proprietary questions and some of the client's questions. The broker analyzes the results for the client, but also keeps the data for its own direct marketing and research applications (National Demographics & Lifestyles, 1992a). Did you ever wonder why those tip-in cards all go to post office boxes in Denver? That is the location of NDL's offices. However, data entry takes place at a facility in Bridgetown, Barbados—offshore data processing and digitizing are relatively new features of the labor geography of GDIS (see Rhind, 1993).

14. Premium selections from NDL's Lifestyle Selector, for example, include households that own camcorders, cellular phones, VCRs, CD players, and home computers; which donate to charities, own stocks or real estate investments, shop by catalog; which are frequent flyers and casino gamblers; and which have incomes over $75,000 (National Demographics & Lifestyles, 1993c).

15. A Grumman advertisement, for example, describes electronic surveillance technologies used for targeting in Operation Desert Storm (Grumman, 1993), while Fairchild compares the management of truck, helicopter, and police car fleets with commanding tanks and naval vessels, promising that its "time-critical" GIS "gives you the technological edge to make rapid-fire, responsible decisions where seconds can mean the difference between successs or failure" (Fairchild, 1993).

REFERENCES

Baker, S., & Baker, K. (1993). *Market mapping: How to use revolutionary new software to find, analyze and keep customers.* New York: McGraw-Hill.

Baudrillard, J. (1981). *For a critique of the political economy of the sign.* (C. Levin, Trans.). St. Louis: Telos Press.

Bondi, L., & Domosh, M. (1992). Other figures in other places: On feminism, postmodernism and geography. *Environment and Planning D: Society and Space, 10,* 199–213.

Burnham, D. (1983). *The rise of the computer state.* New York: Random House.

Cad Images, Inc. (1993). Video tracing. *International GIS Sourcebook 1991–1992.* (p. 383). Fort Collins, CO: GIS World.

Catrell, A. C., Dunn, C. E., & Boyle, P. J. (1991). The relative utility of the Central Postcode Directory and Pinpoint Address Code in applications of geographical information systems. *Environment and Planning A, 23,* 1447–1458.

Claritas. (1993). Prepare to enter a new dimension [Advertisement]. *Business Geographics, 1*(5), 30.

Claritas/National Planning Data Corporation. (1992). *For the demographic state of the nation: Update 92/97* [Brochure]. New York: Author.

Claritas/National Planning Data Corporation. (1993). *Compass: The desktop powertool for precision marketing* [Brochure]. New York: Author.

Claritas/National Planning Data Corporation. (n.d.). *Prizm clusters* [Brochure]. New York: Author.

Curry, D. J. (1993). *The new marketing research systems: How to use strategic database information for better marketing decisions.* New York: Wiley.

de Certeau, M. (1984). *The practice of everyday life* (S. Rendall, Trans.). Berkeley and Los Angeles: University of California Press.

Deetz, S. (1990). Representation of interests and the new communication technologies: Issues in democracy and policy. In M. J. Medhurst, A. Gonzalez, & T. R. Peterson (Eds.), *Communication and the culture of technology* (pp. 43–62). Pullman: Washington State University Press.

Deleuze. G. (1992). Postscript on societies of control. *October, 59,* 3–7.

Digital Equipment Corp. (1993). Everything that's put there can be found right here [Advertisement]. *Business Geographics, 1*(2), 57.

Digital Matrix Services. (1993). GIS—The way it ought to be! InFoCad [Advertisement]. *International GIS Sourcebook, 1991–1992* (p. 370). Fort Collins, CO: GIS World.

Donnelley Marketing Information Services. (1992). *Demograhics on call: The right tool for the information age* [Brochure]. Stamford, CT: Author.

Donnelley Marketing Information Services. (1993). *CONQUEST: Total information power, in one powerful source* [Brochure]. Stamford, CT: Author.

Donnelley Marketing Information Services. (1993b). *Yesterday, a map was a map . . . was a map* [Advertisement]. *Business Geographics, 1*(2), 15.

Dobson, J. E. (1993). Consider both sides of GIS ethics. *GIS World, 6*(9), 20–21.

Equifax Marketing Decisions Systems. (1991). *MicroVision: Geographic consumer targeting system* [Brochure]. Encinitas, CA: Author.

Equifax Marketing Decision Systems. (1993). Introducing INFOMARK-GIS. See your market through a new dimension [Advertisement]. *Business Geographics, 1*(2), 1.

Fairchild Defense. (1993). In the time it takes to read this headline . . . You could be tracking an entire fleet [Advertisement]. *International GIS sourcebook 1991–1992* (p. 479). Fort Collins, CO: GIS World.

Feenberg, A. (1991). *Critical theory of technology.* Oxford: Oxford University Press.

FIND/SVP. (1993). The information catalog. New York: Author.

Fiske, J. (1989). *Reading the popular.* London: Unwin Hyman.

Foucault, M. (1980a). The eye of power. In C. Gordon (Ed. & Trans.), *Power/knowledge: Selected interviews and other writings, 1972–1977.* New York: Pantheon Books.

Foucault, M. (1980b). Questions on geography. In C. Gordon (Ed. & Trans.), *Power/knowledge: Selected interviews and other writings, 1972–1977.* New York: Pantheon Books.

Foucault, M. (1979). Governmentality. *Ideology and Consciousness, 6,* 5–21.

Foucault, M. (1986). *The history of sexuality, Vol. 3: The use of pleasure* (R. Hurley, Trans.). New York: Vintage Books.

Foust, B., & Botts, H. (1993, March–April). Mapping market penetration to discover "hidden" opportunities. *Business Geographics, 1*(2), 38–40.

Francese, P. (1988). How to manage consumer information. In P. Wickham (Ed.), *The insiders guide to demographic know-how* (pp. 10–14). Ithaca, NY: American Demographics Press.

Francese, P., & Piirto, R. (1990). *Capturing customers: How to target the hottest markets of the '90s.* Ithaca, NY: American Demographics.

Francica, J. R. (1992). From frontier to lonely outpost: Businesses move slowly into GIS. *International GIS Sourcebook,* 1991–1992 (pp. 358–359). Fort Collins, CO: GIS World.

Fullerton, R. A. (1985). Segmentation strategies and practices in the 19th-century German book trade: A case study in the development of a major marketing technique. In C. T. Tan & J. N. Sheth (Eds.), *Historical perspectives in consumer research: National and international perspectives.* Proceedings of the Association for Consumer Research, Urbana, IL.

Garvin, A. P. (1993). *The art of being well-informed: What you need to know to gain the winning edge in business.* New York: FIND/SVP.

Graphic Data Systems. (1993). With a clear vision of today, you can paint a clearer picture of tomorrow. *GIS World, 6*(2), 44.

Gelerntner, D. (1989). The metamorphosis of information management. *Scientific American, 261*(2), 66.

Genasys. (1993). With a little advance warning, Pine Canyon wouldn't have become such hot property. *GIS World, 6*(2), 17.

Geoforum. (1993). SMI buys DMIS. *Geoforum, 9*(1), 1.

Geographic Data Technology, Inc. (1993). Their idea of precise geocoding [Advertisement]. *Business Geographics, 1*(5), 56.

GeoQuery Corporation. (1992). *There are many ways to look at your business* [Brochure]. Naperville, IL: Author.

GIS Research Corporation. (1993). Some people take a funny approach to GIS marketing [Advertisement]. *GIS World, 6*(2), 70.

GIS World. (1992). *Discover GIS: The strategic business weapon* [Flier for the "GIS in Business '93" conference, Boston].

GIS World. (1993). Gis World, Inc. announces *GIS in Business '94* conference [Press Release, Auguest 31]. Fort Collins, CO: GIS World.

Goss, J. D. (1993, November 11–14). *We know who you are and we know where you live: The instrumental rationality of Geo-Marketing Information Systems.* Paper presented at the National Center for Geographic Information and Analysis Geographic Information and Society Workshop, Marine Research Center, Friday Harbor, WA.

Grumman. (1993, April 19–25). We read the news before it makes the paper [Advertisement]. *Washington Post National Weekly Edition,* p. 19.

Habermas, J. (1970). *Toward a rational society: Student protest, science and politics* (J. Shapiro, Trans.). Boston: Beacon Press.

Habermas, J. (1973). *Theory and practice* (J. Viertet, Trans.). Boston: Beacon Press.

Haraway, D. (1992). The promises of monsters: A regenerative politics for inappropriate/d others. In L. Grossberg, G. Nelson, & P. A. Treichler (Eds.), *Cultural studies* (pp. 295–337). New York: Routledge.

Harvey, D. (1989). *The condition of postmodernity.* New York: Blackwell.

Holtz, H. (1992). *Databased marketing.* New York: Wiley.

Horkheimer, M., & Adorno, T. (1972). *The dialectic of the enlightenment.* New York: Seabury Press.

Huff, D. (1993). Regional shopping center frenzy prompted Huff model (Interview with David Huff). *Business Geographics, 1*(5), 38.

Hughes, A. M. (1991). *The complete database marketer: Tapping your customer base to maximize sales and increase profits.* Chicago: Probus.

Kling, R. (1985). Computers and social power. In D. G. Johnson, & J. W. Snapper (Eds.), *Ethical issues in the use of computers* (pp. 63–75). London: Wadsworth.

Lake, R. W. (1993). Planning and applied geography: Positivism, ethics, and geographic information systems. *Progress in Human Geography, 17*(3), 404–413.

Lakoff, G. L., & Johnson, M. (1980). *Metaphors we live by.* Chicago: University of Chicago Press.

Lash, S., & Urry, J. (1987). *The end of organized capitalism.* Madison: University of Wisconsin Press.

Lefebvre, H. (1991). *The production of space* (D. Nicholson-Smith, Trans.). New York: Blackwell.

Leiss, W., Kline, S., & Jhally, S. (1986). *Social communication in advertising: Persons, products and images of well-being.* New York: Methuen.

MapInfo. (1993). Hidden plots discovered: Desktop mapping held responsible [Advertisement]. *Business Geographics, 1*(5), 2.

MapInfo. (n.d.). *Introducing MapInfo 2.0: The most advanced MapInfo ever* [Brochure]. Troy, NY: Mapping Information System Corporation.

Marcuse, H. (1964). *One-dimensional man: Studies in the ideology of advanced industrial society.* Boston: Beacon Press.

Martin, D. (1991). *Geographic information systems and their socio-economic applications.* New York: Routledge.

Moloney, T. (1993). Manufacturing and packaged goods. In G. H. Castle (Ed.), *Profiting from a geographic information system* (pp. 105–128). Fort Collins, CO: GIS World.

Morris, M. (1988). Things to do with shopping centres. In S. Sheridan (Ed.), *Grafts: Feminist cultural criticism* (pp. 194–225). New York: Verso.

Morris, M. (1992). On the beach. In L. Grossberg, G. Nelson, & P. A. Treichler (Eds.), *Cultural studies* (pp. 450–478). New York: Routledge.

Naisbitt, J., & Aburdene, J. (1990). *Megatrends.* New York: William Morrow.

National Demographics & Lifestyles. (1991). *Z: Zip screens* [Brochure]. Denver, CO: Author.

National Demographics & Lifestyles. (1992a). *Communicating directly with your consumers* [Brochure]. Denver, CO: Author.

National Demographics & Lifestyles. (1992b) *Sometimes it's difficult to know the territory without a Scout* [Brochure]. Denver, CO: Author.

National Demographics & Lifestyles. (1993a). *Customer database development program. Database marketing services.* Denver, CO: Author.

National Demographics & Lifestyles. (1993b). *NDL FOCUS: Knowing your customers* [Brochure]. Denver, CO: Author.

National Demographics & Lifestyles. (1993c). *The lifestyle selector: The list that lasts* [Brochure]. Denver, CO: Author.

National Demographics & Lifestyles. (1993d). *To know your customers better, get all the details in focus . . .* [Brochure]. Denver, CO: Author.

National Demographics & Lifestyles. (n.d.). *Character. Not just characteristics* [Brochure]. Denver, CO: Author.

National Decision Systems. (1993). *Data packages.* Equifax Marketing Decision Systems. Encinitas, CA: Authór.

Openshaw, S., & Taylor, P. J. (1981). The modifiable unit area problem. In N. Wrigley & R. J. Bennett (Eds.), *Quantitative geography: A British view* (pp. 60–69). London: Routledge & Kegan Paul.

Ozanne, J. L., & Stern, B. B. (1993). The feminine imagination and social change: Four feminist approaches to social problems. In L. McAlister & M. L. Rothschild (Eds.), *Advances in consumer research* (Vol. 20, p. 35). Provo, UT: American Marketing Association.

Pickles, J. (1991). Geography, GIS, and the surveillant society. *Papers and Proceedings of the Applied Geography Conferences, 14,* 80–91.

Piirto, R. (1991). *Beyond mind games: The marketing power of psychographics.* Ithaca, NY: American Demographics.

Poster, M. (1990). Foucault and databases. *Discourse, 12*(2), 112–127.

Ramtek. (1993). For a better view of the terrain. *International GIS Sourcebook 1991–1992* (p. 600). Fort Collins, CO: GIS World.

Reisberg, G., & Gilbert, S. (1991, February). Finding quality propects: New techniques let you make the most of your business database. *Target Marketing,* pp. 59–61.

Rhind, D. (1992a). The next generation of geographical information systems and the context in which they will operate. *Computer Environments and Urban Systems, 16,* 261–268.

Rhind, D. (1992b). War and peace: GIS data as a commodity. *GIS Europe, 1*(8), 24–26.

Ries, A., & Trout, J. (1986). *Marketing warfare.* New York: McGraw-Hill.

Roberts, K. (1992, April). Aiming for a direct hit. *Information Week,* pp. 26, 30.

Roszak, T. (1986). *The cult of information: The folklore of computers and the true art of thinking.* New York: Pantheon Books.

Slater, D. (1992). Going shopping: Markets, crowds and consumption. In S. Lash & J. Friedman (Eds.), *Modernity and identity* (pp. 188–209). Cambridge, MA: Blackwell.

Smith, N. (1992). Real wars, theory wars. *Progress in Human Geography, 16*(2), 257–271.

Stoecker, D. (1993). Attribute data for GIS. In G. H. Castle (Ed.), *Profiting from a geographic information system* (pp. 231–255). Fort Collins, CO: GIS World.

Strategic Mapping, Inc. (1993). Secret weapon [Advertisement]. *Business Geographics, 1*(5), 4.

Tactics International Ltd. (1991). *Tactician: International micro marketing solutions* [Brochure]. Andover, MA: Author.

Tactics International Ltd. (1993). *Tactician: International micro marketing solutions* [Brochure]. Andover, MA: Author.

Tactics International Ltd. (n.d.). *Tactician becomes the selling machine and the micro-marketing machine* [Brochure]. Andover, MA: Author.

Tetzeli, R. (1993, October). Mapping for dollars. *Fortune,* pp. 90–95.

Thomas, R. K., & Kirchner, R. J. (1991). *Desktop marketing: Lessons from America's best.* Ithaca, NY: American Demographics Books.

Toffler, A. (1980). *The third wave.* New York: William Morrow.

Tosta, N. (1993, February). The data wars, part 2. *Geo Info Systems,* pp. 22–26.

Tydac Technologies Corp. (1993). Discover the hidden worlds in your data [Advertisement]. *Business Geographics, 1*(2), 12.

Urban Decision Systems. (1992). *Maps on call* [Brochure]. Los Angeles: Author.

Urban Decision Systems. (n.d.). *Visualize your accurate, believable, comprehensive data* [Brochure]. Los Angeles: Author.

Veblen, T. (1954). *The theory of the leisure class.* New York: Mentor Books. (Original work published 1899)

Virilio, P. (1991). *The aesthetics of disappearance* (Philip Beitchman, Trans.) New York: Semiotext(e).

Weiss, M. J. (1988). *The clustering of America.* Cambridge, MA: Harper and Row.

Winner, L. (1986). *The whale and the reactor: A search for limits in an age of high technology.* Chicago: University of Chicago Press.

Wolf, A. (1990). How to fit geo-objects into databases: An extensibility approach. *Proceedings: European Geographic Information Systems* (Vol. 2, p. 1164). Amsterdam, the Netherlands.

CHAPTER 8

Earth Shattering

GLOBAL IMAGERY AND GIS

Susan M. Roberts
Richard H. Schein

Geographical information systems (GIS), global positioning systems (GPS), computer aided drafting (CAD), and a whole host of variations on computer systems designed to process cartographic and spatial data have been major growth industries for the past 20 years. As these industries have boomed, a range of periodicals, trade magazines, and newsletters have been established to serve as forums for the display and discussion of innovative approaches, new tools, and applications. This chapter focuses on one aspect of this burgeoning print media. Using concepts from Lefebvre (1974), we critically examine magazine advertisements for geographic technologies as representations that may be used to interrogate the relations between spatial data-based technologies and the spatial practices of contemporary society. Advertisements are concise statements designed to convey vendors' messages about their products. The size restrictions on these advertisements—they rarely fill more than one page and are often smaller than a single page—combined with the competition between them to secure the reader's attention have given rise to clever and eye-catching uses of pictorial imagery. Pictorial imagery compresses the advertisers' claims into a visual "shorthand" that interacts with and reinforces the advertisement's text. In this chapter we are particularly concerned with shorthand, visual messages that employ images of the globe and the earth's surface. The multitude of ways in which those particular images are incorporated into advertisements tells us much about the underlying relations between the new technologies and their subject: the world.

Lefebvre's (1974) analysis of the social production of space provides a "conceptual triad" for teasing out the relations between GIS (and its associated technologies), the advertisements for those technologies, and the spatiality of contemporary life. According to Lefebvre, every social formation secretes, or produces, its own space. That process is dialectical and reflexive, and has three identifiable moments:

1. Through *spatial practices* each society secretes its own distinctive material spatial arrangements. *Spatial practices* include production and reproduction, and entail the daily routines and flows that secrete communications and transport networks, produce urban hierarchies, and differentiate public and private spaces.
2. *Representations of space* are conceptualizations of ordered space: the space of scientists, engineers, planners, and technocrats. *Representations of space* are the dominant space in any society, and it is through the system of signs or codes upon which they are based that individuals *know* space and its ordering. A map, chart, or blueprint is a *representation of space.*
3. *Representational spaces,* on the other hand, are more often systems of nonverbal symbols and signs. *Representational spaces* are dominated spaces, appropriated by the imagination in order to describe through symbols and images. Works of artists, photographers, and filmmakers may be *representational spaces* (Lefebvre, 1974, pp. 1–67; see also Harvey, 1989, pp. 256–278).

In this chapter, print advertisements—particularly those employing pictures of the globe—are treated as *representational spaces.* Through the use of symbols and images (of the globe and the earth's surface) advertisements seek to summarize and distinguish particular *representations of space* (such as a GIS). *Representational spaces* (advertisements) and *representations of space* (GIS and associated technologies) are wrapped up in, implicated in, society's *spatial practices,* or our daily experienced material reality. Each element of Lefebvre's triad, *representational spaces, representation of space,* and *spatial practices,* (or perceived, conceived, lived) is interconnected and mutually constitutive, in coherent or disjunctive ways (Lefebvre, 1974,

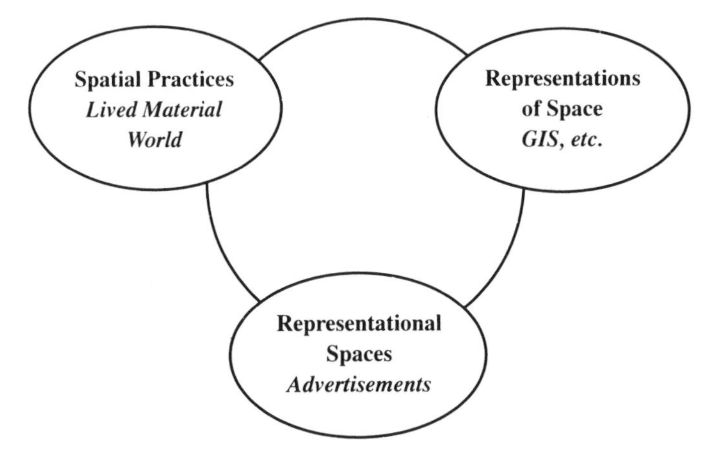

FIGURE 8.1. Lefebvre's triad.

p. 42). The triadic relationship examined in this chapter is shown in Figure 8.1.

In this chapter we do not pretend to offer an exhaustive or comprehensive content analysis of print advertisements for geographic technologies. However, by specifically focusing upon the nature of print advertisements as *representational spaces,* we will be able to reflect upon their dialectical relationships with the technologies they are selling (*representations of space*) and the world we inhabit (*spatial practices*). The second part of this chapter begins with an examination of recent print advertisements as *representational spaces.* The third part unpacks the geographic technologies presented in the ads as ways of seeing or ordering the world: as *representations of space.* The fourth part examines the implications of the technologies and their advertisements in the contemporary world, or as part and parcel of society's *spatial practices.*

ADVERTISEMENTS AS REPRESENTATIONAL SPACES

Advertisements for geographic technologies are *representational spaces* using symbols and images in order to concisely convey the specific technology's capabilities. The overwhelming message of these advertisements concerns the ability afforded the user of particular technologies to capture the whole world, and to reduce it to technologically manipulatable components.

First, the presence of the globe or a map in these ads speaks to more than simply the inherent cartographic function of GIS and other technologies. The globe symbolizes the extent of coverage possible with technological innovation. Quite literally, the earth is the limit. Depicting the world from above not only suggests the sophisticated, space-age equipment used in, say, a global positioning system; it also claims the entire world as a potential data source. Depictions of the globe place the world at one's fingertips (Genasys, 1990) and position the viewer high above the earth to claim a view that is total (see Figures 8.2 and 8.3) (see de Certeau, 1984; and Barthes, 1979, on views from above). Views from above underscore the advertiser's totalizing claims of complete coverage, of being everywhere.

FIGURE 8.2. From Erdas (1991, p. 7). Copyright 1991 Erdas. Reprinted by permission.

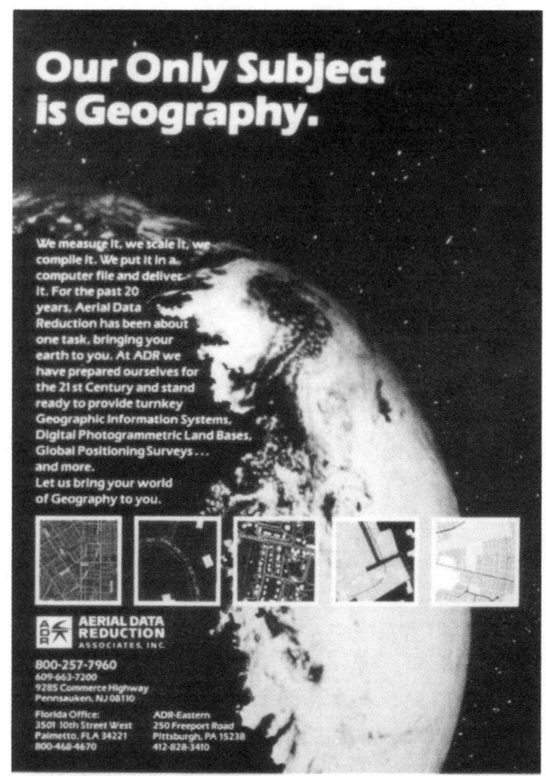

FIGURE 8.3. From Aerial Data Reduction Associates, Inc. (1990, p. 36). Copyright 1990 Aerial Data Reduction Associates, Inc. Reprinted by permission.

Second, by juxtaposing such a view of the world with the image of a floppy diskette or of a computer screen, the advertisement establishes the capability of the technology to reduce the world to fit into a personal computer (PC) (see Figures 8.2 and 8.4). With a PC the world can be captured or, as one advertisement boldly asserts, "everything that's out there can be found right here" (see Figure 8.5). Several advertisements depict a single computer screen that frames a portion of a larger map or globe image behind the machine (see Figures 8.2 and 8.6). Similarly, aerial photographs, remotely sensed imagery, or more conventional cartographic data are often projected from an adjacent computer screen, implying that it is only a matter of scale manipulation to move from the "real world" to the world in the PC.

FIGURE 8.4. From Strategic Mapping, Inc. (1991, p. 11). Copyright 1991 Strategic Mapping, Inc. Reprinted by permission.

Capturing the world within the computer is like buying or owning a piece, or potentially all, of the earth. With part of the world inside the computer, the technologically sophisticated user is in a position to control, manage, or otherwise manipulate not only the data, but by inference, the world itself (since that is what is "inside" the computer). Many new ads for PCs depict the computer as a souped-up race car, in effect putting the computer operator firmly in control, behind the wheel of data and earth-space manipulation (see Figures 8.7 and 8.8).

In GIS, in particular, the utility of the hardware/software combination is found in the user's ability to "integrate" or "stack" data. A GIS is, after all, fundamentally a series of overlain maps and linked databases. One data vendor's advertisement presents a picture of a multilayered sandwich, waggishly suggesting that by buying the product "your GIS" will be "a hero" (SPOT Image Corporation, 1992). Many ads stress the relationship between the "whole picture" and the "pieces" (see Figure 8.9) that are to be integrated, ideally in a "seamless" (see Figure 8.10) re-creation of a real-world totality. Advertisements that stress the ability of a GIS to integrate many different spatial data sets with the immediate goal of better depicting the world are similar to those that simply present the globe as the focus of a total technological view. The ultimate (supposed) aim of the advertiser is to stress the ability of particular machines, software, and data sets to provide the user with ever more complete and accurate windows on the real world. As the ad for one GPS proudly suggests, "our customers routinely ask the world of us" (Trimble, 1990). In marketing strate-

FIGURE 8.5. From Digital Equipment Corp. (1992, p. 67). Copyright 1992 Digital Equipment Corp. Reprinted by permission.

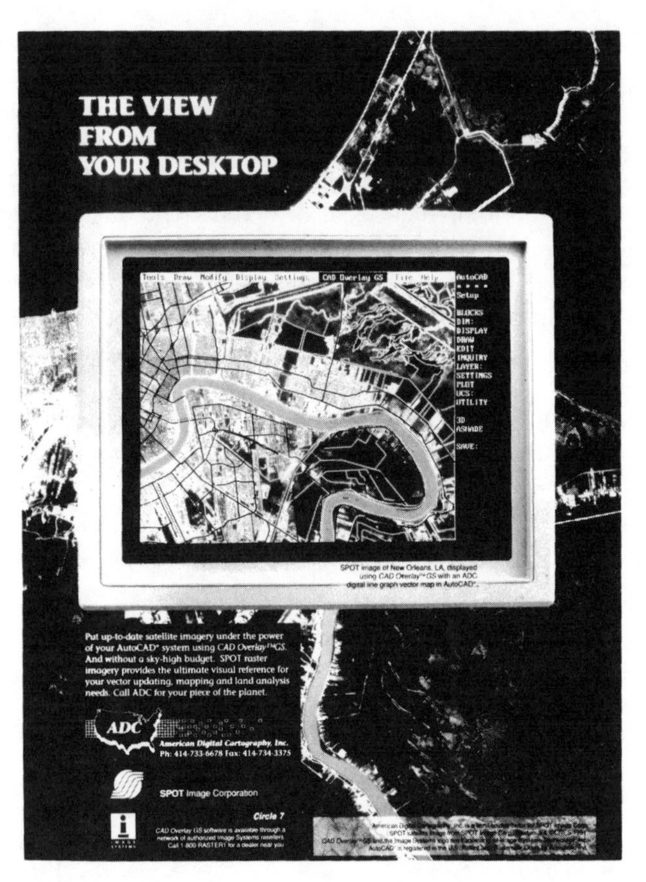

FIGURE 8.6. From American Digital Cartography, Inc. (1991, p. 9). Copyright American Digital Cartography, Inc. Reprinted by permission.

gies that portray seemingly limitless opportunities to capture, manipulate, and portray the earth or parts of it, we might note the Orwellian claim of Autodesk, Inc. (1992), who, presenting us with a picture of a typical American suburban neighborhood complete with family dog, proudly proclaim that "there's only one thing in this picture that isn't managed by autocad."

GEOGRAPHIC TECHNOLOGIES AS REPRESENTATIONS OF SPACE

The fundamental assumption of any GIS technology (and its advertisements) is that the user can *reproduce* real-world space—exact *copies* of that which normally meets the eye. Or, even better, the entire picture does not have to be taken. Instead, the user can isolate pieces of the real world, or separate them into thematic overlays that will eventually be put back together to reproduce the whole picture. In both cases, the technology promises a mimetic reproduction of the visual, a representation of earth space that is simply and unproblematically a reflection of what is really "out there."

Behind the fundamental assumption, however, is a several-hundred-year history of Western concepts of space, vision, and representation that bears directly on the presentation and use of GIS and its associated technologies as abstract ways of ordering, seeing, and knowing the world. GIS and allied systems are culturally embed-

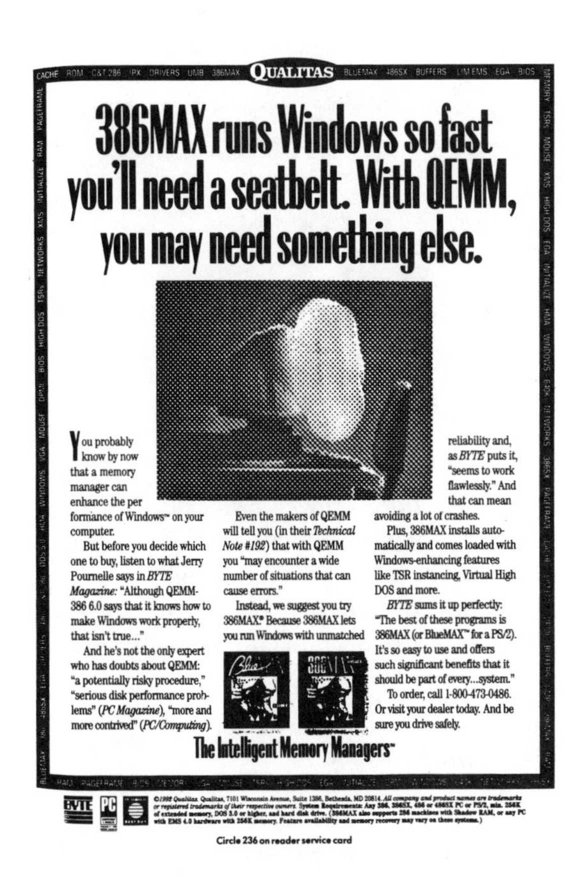

FIGURE 8.7. From Qualitas (1990, p. 67). Copyright 1992 Qualitas. Reprinted by permission.

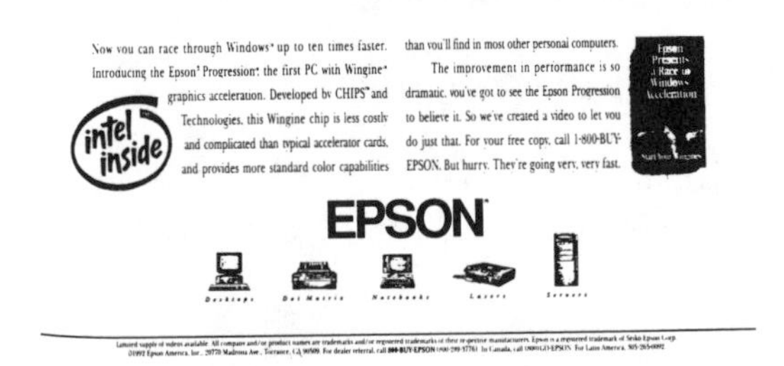

FIGURE 8.8. From Epson America Inc. (1992, p. 57). Copyright 1992 Epson America Inc. Reprinted by permission.

ded *representations of space*—they are socially produced ways of seeing. Geographic technologies as ways of seeing also are embedded materially in the *spatial practices* of a particular political economy, that of late capitalism. Geographic technologies are very much a part of the ways in which particular social formations secrete distinctive spaces.

Geographic Technologies: Ways of Seeing

The space that is being reproduced in a GIS simulation is an abstract Cartesian space, a three-dimensional reality that is reduced in scale to "fit" onto the computer screen. The ways in which that space is viewed (and the computer screen itself) are enabled by constructed rules of vision that allow us to separate ourselves from the image as the object of our investigations. But such GIS images are not simply unproblematic reflections of the world. Although claiming that "we can change the way you see the world" (Radian Corp., 1992), GIS produce representations that draw upon the invented Western traditions of three-dimensional space, subject–object relations, and perspective (see Cosgrove, 1988; Crary, 1990; Mitchell, 1986). The image on the computer screen is embedded within those taken-for-granted social ways of knowing, or *epistemes.* The image or representation is a constructed abstraction, although not simply in the conventional cartographic sense of abstracting a piece of reality. Rather, a GIS (for example) offers selec-

FIGURE 8.9. From EDS. (1992, p. 17). Copyright 1992 EDS. Reprinted by permission.

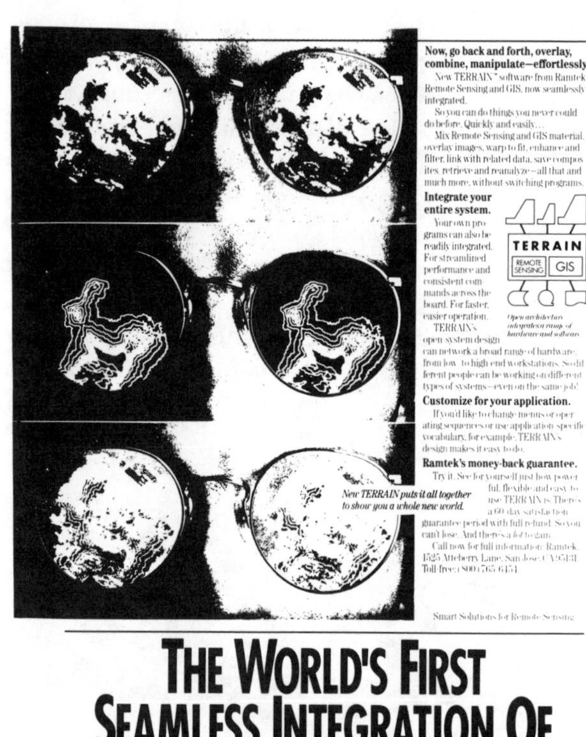

FIGURE 8.10. From Ramtek (1991, p. 17). Copyright 1991 Ramtek. Reprinted by permission.

tive images. These selected *representations of space* exist within particular frames of knowing. Thus to read or interpret the advertisements—and by implication, the messages and utilities of GIS, GPS, remote sensing, and other related technologies—it is useful to realize that the images presented to us are not mimetic reproductions or transparent windows on a naively given real world. They are constructed images created within constructed rules of vision, representation, and space. We can interrogate the images and technologies not with an attention simply to their ability to "accurately" reflect a portion of the earth's surface (in terms of resolution, scale, scope of coverage, etc.), but in order to ask questions about the messages they contain regarding the sociospatial world, our interpretation of it, and our ongoing reproduction of it through spatial practices.

Vision is privileged as a way of knowing. That which is seen is real. An ad for Ramtek that presents computerized images of the earth's surface reflected in several pairs of eyeglasses suggests that what you see is what you get (see Figure 8.10). Of course, no one actually *sees* exactly the remotely sensed data presented in the ad. Yet it is understood that the view presented is *seen* by the satellite which in turn enables the reader's view. When we view and, importantly, when we record a view, especially a view from above, we are establishing our own superiority and our domination of the scene. We look for that which we recognize, and it becomes the object of our investigation. We imagine ourselves as separate from the view, situated somehow outside the data, and the view or its contents become ours to control and manage.

This proprietorial objectivism is certainly the message of many contemporary advertisements. GIS and related technologies promise to literally enframe a view of the earth's surface for our use and control. Thus the message in the claim that "everything that's out there can be found right here" (see Figure 8.5) not only gives the user access to (supposedly) all the information in the world, but that view of the world becomes "the view from your desktop" (see Figure 8.6). Or, as one advertisement tantalizingly suggests, "Imagine buying all this real estate for only $99!" (Horizons Technology, Inc., 1992). The technology allows "imagined proprietorship" of a selected part of the world (Trachtenberg, 1991, p. 42).

Once a part of the world is "in" the computer, the users can do with it what they please. To extrapolate from several recent advertisements for PCs, the amount of information available today puts the user in the driver's seat (see Figure 8.7). PC ads seem to have co-opted the symbolism of the automobile in appealing to the consumer. There is much symbolic baggage associated with automobile imagery; after all, the automobile is an integral part of American life and has been for at least 80 years. Questions are raised about the appeals to machismo, for example, the masculine orientation of the big engine, racing stripes, speed, and the emphasis on "performance" (see Figure 8.8). More specifically, there are definite spatial implications of adopting automobile metaphors for computer technologies.

The automobile, or more precisely affordable mass-produced automobiles, were certainly the 20th century's most significant technological innovation vis-à-vis the alignment and realignment of the *spatial practices* of day-to-day lives in North America at least. There is a voluminous literature in which the automobile is examined as the epitome of individual freedom and spatial mobility (see, e.g., Bayley, 1986; Flink, 1975; Rae, 1971). By borrowing automobile images, PC ads are not simply appealing to (primarily masculine) consumer tastes in familiar, time-tested ways. They are also making implicit statements about the power of the computer to surpass and supplant the automobile as technological innovation.

The automobile revolutionized the *spatial practices* and constructs of everyday lives: the American urban shift to decentralized, or galactic, cities begun in the late 19th century reached its apogee thanks in part to the interstate highway system and personal motor-car use in the 1960s. The suburb is one of the most potent symbolic landscapes of late-20th-century America, and its existence and imagery are highly dependent upon the automobile (Meinig, 1979). In short, the automobile was the *Fordist* space-maker in the American landscape.

The PC, by analogy, is the quintessential *post-Fordist* space-maker. GIS operated on desktop computers of which the user is the driver allow the juxtaposition of images and codes unseen in the "real world" of our daily lives. Space becomes layered and fragmented (see Figure 8.9; SPOT Image Corporation, 1992). The technological sophisticate with the correct machines and software can align and realign spatial images in many ways. GIS packages allow a reconfiguration of the "real world" captured inside the computer, perhaps ultimately dispensing with the "real world" altogether, in favor of a humanly created cyberspace, existing only inside the computer with no reference to material spatial practices beyond the microprocessor.

GIS work as *representations of space* in reconfiguring and reordering spatial elements in large part through their ability to concretize and naturalize produced images as mimetic windows on the world. When a GIS is employed, the user determines the parameters of the data overlays; but having done so, the user may be tempted to ignore the fact that the images received and manipulated within those overlays are constructed. What you see *may be* what you get; but what is seen is far from an unproblematic reflection. GIS is a *representation of space* offering conceptualized space, based on, but not the same as, that space which a society secretes through its material spatial practices. Furthermore, a GIS does not present a representation of a world divorced from its investigation. A GIS offers pictures of a world that has been socially created and is in the process of being socially interpreted. The GIS package, the computer, and hard-copy images are part and parcel of human *spatial practices.* The applied technologies are implicated in a reflexive relationship in which the world is viewed, captured, interpreted, presented, and, finally, viewed again. Thus the images produced, the spaces portrayed, and the messages of the technologies, their creators, and their users are all part of the instrumental ordering of our sociospatial world(s) and of our lived *spatial practices.*

The Political Economy of Geographic Technologies

GIS and its associated technologies are *representations of space:* ways of seeing that order the world within a familiar Cartesian framework. As especially technical or mechanistic ways of seeing, they also are embedded in a particular political–economic context. That context includes the contemporary geographies of uneven development and their associated asymmetries of power. Furthermore, a discussion of that context draws the explication of geographic technologies, as *representations of space,* and their ads, as *representational spaces,* closer to the third moment of Lefebvre's triad, *spatial practices.*

Geographic technologies and their global imagery represent the world from above in a view that is possible only with space flight (see Figures 8.2 and 8.3). Earlier attempts at aerial photography and reconnaissance, using pigeons, balloons, and planes, could give views of parts of the earth's surface, but a view of the earth as a whole spinning solitarily in space could only come with space flight (space photography) and satellite technologies (remote sensing) (Barry, 1992). The view of the earth as a whole is thus that of the space*man* who has actually attained the god's view—a perspective only imagined by cartographers from the Renaissance onward (Cosgrove, 1989, p. 15). The acquisition of the god's view is invoked playfully by the GPS advertisement that advises: "Take stock of your earthly positions, with some guidance from above" (Canadian Marconi Co., 1992). The image of the satellite as divine is a recent expression combining the long-standing connection between the "authority" attributed to the camera lens with a god's omnipresent and unerring observation of all of human life (see Trachtenberg, 1991, p. 18). In her discussion of the effects of "visualizing technologies" Haraway (1991, p. 189) puts it this way: "Vision in this technological feast becomes unregulated gluttony; all perspective gives way to infinitely mobile vision, which no longer seems just mythically about the god-trick of seeing everything from nowhere, but to have put the myth into ordinary practice."

The technologies that enable the view from space are those of space flight and satellites. Space views of earth may be used to represent the earth as the seamless home of humankind, but it is only the inhabitants of a certain portion of the earth's surface who have developed and applied space flight and satellite technologies. Specifically, the governments of the United States, the former Soviet Union, and European countries (notably France) have, since the end of World War II, devoted substantial resources to the growth of nationally defined space programs. Such programs have been at the heart of the ongoing construction of national identities. During the cold war the intense competition between the United States and the USSR to claim "firsts" drove the development of space technologies. Such technologies commonly had both military and commercial applications. For instance, the rocket technology developed for launching intercontinental ballistic missiles (ICBMs), keystone geostrategic weaponry in the cold war, and the rockets for launching satellites are very similar. The view of the earth from space, in its origins, is thus embedded in the spatial practices of national military programs defined by, and themselves defining, the cold war (McLucas, 1991, pp. 9–10). The "New World Order" may not be so different: SPOT images were employed in Operation Desert Storm (Barry, 1992, p. 571), and Schulman (1991, pp. 28–29) suggests that the use of advanced GIS and related technologies in the Persian Gulf War "remove[d] a lot of combat anxiety, . . . substantially increased the mission's effectiveness and contributed to the safety of military personnel."

As the United States government was pursuing the geopolitical goal of hegemonic militarism, American corporations were recasting the previous patterns of the world economy through a geopolitics of their own. American corporations (especially those in the motor vehicle and oil industries) pursued growth policies through aggressive multinational investment and sales. By 1973 General Motors's annual sales were bigger than the gross national product of either Switzerland, Pakistan, or South Africa (Barnet & Müller, 1974, p. 15). During the cold war, American multinationals grew big, powerful, and global. Barnet and Müller (1974) argued that "the most revolutionary aspect of the planetary enterprise is not its size but its *worldview*" (p. 15, emphasis added). This worldview was global. To be sure, the American multinationals of the cold war era had precursors in the large and powerful colonial trading companies, but the multinationals' technologically enabled claims of scale, scope, and especially of a supra- or anational identity were new. As Barnet and Muller (1974) stated:

> Under the threat of intercontinental rocketry and the global ecological crisis that hangs over all air-breathing creatures, the logic of global planning has become irresistible. . . . Rising out of the post-World War II technological explosion which has transformed man's view of time, space, and scale, global corporations are making a bid for political acceptance beyond anything ever accorded a business organization. The first entrepreneurial class with the practical potential to operate a planetary enterprise now aspires to become global managers. (p. 14)

While in some cases the activities of cold war militarism and the rise of the American multinational went hand in hand (ITT being a notorious example), both political and economic activities were supported by technology-enabled *representations of* global *space.* American geopolitical and management strategies held to a global worldview.

The military origins of the technologies enabling the earth to be seen from space alert us to the role of the state as a central actor in the story of GIS. Today, many GIS applications draw on data collected from satellite systems set up by arms of the state. Two of the best known examples are the U.S. Landsat and the French SPOT satellite systems. These operate remote sensing satellites that orbit the earth every 16 and 26 days, respectively, collecting data through electromagnetic radiation. This venture out of the visual part of the electromagnetic spectrum (e.g., into the infrared part) enables data to be collected and then digitally re-presented visually as an image that looks like a photograph or even a map, but in origins is neither. The data obtained by these satellites is transmitted instantaneously down to receiving stations on earth (Gardner, 1992, p. 153). Remotely sensed data from satellites may be the basis for a GIS or may be combined with data collected by other means (see Figures 8.6 and 8.10). In the early days of NASA's Landsat development, American oil and mining companies pioneered sophisticated techniques to use Landsat data for oil and minerals exploration in the United States and around the world. For example, Mack (1990, p. 175) reports that in 1975 "Chevron Oil claimed to be spending $50,000 to $100,000 per year on obtaining and interpreting Landsat data." The use of representations of the earth as seen from space as well as the actual employment of remotely sensed data in GIS indicate *spatial practices* characteristic of the cold war political economy. Significantly, these *spatial practices* rested in part on technology-enabled *representations of* global *space.*

These *spatial practices* entailed increasing mobility. In a GIS the view is not static, and neither is the viewer/user fixed in relation to the view. This is a defining feature of GIS and its effects may be compared to those of the movie camera. When the movie camera was new, its new way of seeing prompted Walter Benjamin (1936/1968, p. 228) to write: "Guided by the cameraman, the camera continually changes its position with respect to the performance." This mobility of the viewer (the GIS user) relative to the earth (as represented in a set of spatial data) echoes the comments made by Soviet film director Dziga Vertov (quoted in Berger, 1972, p. 17) in 1923: "I'm an eye. I, the machine, show you a world the only way I can see it. I free myself for today and forever from the human immobility. I'm in constant movement. . . . Freed from the boundaries of time and space, I co-ordinate any and all points of the universe, wherever I want them to be." In the words of one advertisement, "the perfect 3-D visualization tool for users of Landsat, SPOT, other digital imagery, and GIS systems!" offers "near real-time fly through of imagery draped over terrain; interactive control of position, direction, elevation, ground-speed, and look direction; interactive control of exaggeration, view cone, horizon tilt, zoom factor" (PCI Remote Sensing Corp., 1992).

This mobility of the viewer adds to his or her power over the subject matter, allowing a thorough inspection from all angles. The ability to have one's subject matter so completely open to inspection and examination depends on an indifference toward the inhabitants whose *spatial practices* are captured as spatial data. The distance between the earth's surface and the satellite allows people to be unaware that their picture is being taken, or if aware and unwilling, unable to decline such attention. From 1968 onward the United Nations has attempted to set up international legal arrangements for the collection and use of satellite-gathered data. The United Nations has found this to be a contested and politically charged issue, with countries disputing the rights of other countries to obtain, firsthand or otherwise, satellite-gathered data regarding their own territory. For example, in 1971 Mexico "made it clear to the United Nations that it expected 'no data would be collected over Mexican territory from the air or space without prior permission' " (Mack, 1990, p. 187).

The satellites' way of seeing is that of the voyeur or even of the violator. The sexual imagery is deliberate: A GIS is a gendered technology relying on scientific knowledge; it is the product of a scientific mind conceived as male and disembodied (Keller, 1992, p. 19). The technology is socially constructed as masculine in the same way that the camera itself has been recognized as an extension of a "redoubtable masculine will" implying (or forcing) the subject's "surrender" (Trachtenberg, 1991, p. 27). In the case of GIS and related technologies, the *representation of space* through the collection, inspection, and manipulation of spatial information is clearly the inheritor of the cartographic tradition of seeing. Mapping, locating, and differentiating territory was a project of the modern state and went hand in hand with the discovery, penetration, conquest, pacification, and possession of vast swathes of the earth (see Driver, 1992; Harvey, 1989; Livingstone, 1992). Just as cartography and navigation combined to make the world known to Europeans, they also made it prey to Europe's greed. Events described as "the imperial conquest of virgin lands" or "the penetration of darkest Africa" served to reinforce the identification of space as feminine. Thus, William Blake (in "A Vision of the Last Judgment") stated: "Time & Space are Real Beings. . . . Time is a Man, Space is a Woman. . . ." The gendered Cartesian conception of space as inherently feminine and passive has been a corollary of the dominance by the temporal as the domain of active masculine history making—the so-called annihilation of space by time (for critiques see, e.g., Soja, 1990; Massey, 1992). Passive space is implied in the transformation of the earth's surface into spatial data. Socially produced space becomes categorized spatial data and its recombinations: It becomes "an image and the reproductions of an image" that allow it to be possessed (Benjamin, 1931/1980, p. 209). Through inspection and viewing as spatial data, space is thoroughly possessed as "real estate" (Horizons Technology, Inc., 1992). "Call ADC for your piece of the planet" (see Figure 8.6) or "Let us bring your world of Geography to you" (see Figure 8.3). Trachtenberg (1991, p. 42) has noted that "imagined proprietorship" through the consumption of images of the world invites "fantasies of power." Such fantasies are gendered—most obviously in the high-performance automobile imagery. What Haraway (1991, p. 172) calls the "informatics of domination" is played out in GIS in an active way, as the GIS user takes such images, ruptures and cuts them, and recombines the elements in whatever form he wishes. This violent disruption of a totality ought not to be surprising since scientific discourses and the practice of science have been shown to be associated actively with violence—especially that violence perpetrated by the state (Nandy, 1988; Alvares, 1992; Giddens, 1985, pp. 236–254).

The point of GIS is to "handle" spatial data in a variety of ways. A GIS is deliberately designed to perform a number of operations: to "capture; edit; structure; restructure; manipulate; search; analyze; and integrate" data (Raper, 1992, p. 172). All this may be achieved without waiting: "Everything you need appears on screen in seconds, ready for further interrogation or manipulation" (see Figure 8.9). In this totalization, nothing is hidden from, or immune to, this "handling" where the data is "manipulated" by computer application rather than by the direct touch of the hand (Latin, *manus*) of the user. The resultant representations of space appear disrupted: a sort of dynamic, even violent, shifting collage of fractured images (Harvey, 1989, p. 302; see also Lefebvre, 1974, p. 358).

SPATIAL PRACTICES

The manner in which spatial information is treated in GIS reflects the application of GIS to commercial and public projects. The commercial uses of GIS are numerous. For example, remotely sensed data may be used to "lead real estate developers to properties that contain a minimum of ecologically sensitive wetlands" (EOSAT, 1991). A GIS is a *representation of space* of practical interest not only to real estate developers. Images of the earth used in advertising GIS are reminders of the extent to which capital has globalized, and of the ways in which images of global space have become crucial to the representation of both GIS and multinational corporations. In the contemporary world economy globalization may be seen as highly contradictory processes in which the imperatives of spatial scope and coverage entail a need to amass and process details about the particularities of place. The "earth-shattering" (Intergraph, 1992) capabilities of GIS fit neatly with the need for detail exemplified in the Zip Code classifications of household consumption profiles for the direct mail industry and in the use of detailed consumption data in corporate planning regarding customizing or differentiating a standard product for identified niche markets. For example, one advertisement (Digital Matrix Services, 1991) notes the product's ability to perform "Address geo-coding and matching."

Even as capital works to secure a more homogeneous, friction-free global space, the "diminishing spatial barriers give capitalists the power to exploit minute spatial differentiations to good effect" and make them "more sensitized to what the world's spaces contain" (Harvey, 1989, p. 294). This is a major facet of the "demand side" of GIS and associated technologies. The ability to acquire vast amounts of detailed spatial information, on nearly every corner of the world, is the entirely appropriate technology for capital in an era of flexible accumulation. The imperative for capitalists to at once think both globally *and* locally is catered to. In addition, the world economy has become increasingly characterized by systemic risk as capital has become more mobile, and as the major elements of the world financial system (inflation, exchange rates, and interest rates) have become extremely volatile. In order to turn the risks of uncertainty into opportunity, capitalists seek detailed information. A GIS is just one type of information technology that can be deployed in the profitable negotiation of risk in the world economy (see Hepworth, 1990; Thrift, 1990).

The local state is a major purchaser of GIS, and several advertisements we examined are aimed at this audience. Land use planning, political redistricting, environmental management, and policing are salient applications. For example, U.S. Bureau of the Census data may be used to develop systems of rapid response, to identify crime-prone areas, and to "optimize patrol allocation" (Strategic Mapping, Inc., 1991). Like maps before, GIS is bound up in the changing nature of the state. Knowledge of territorial content within the borders may be used to allocate resources or to supervise and discipline social activity (see Giddens, 1985; Foucault, 1977). Knowledge of the spaces contained by others' borders enables strategic military and economic planning and action. The role of information in managing risk for capital is connected to the role of information in interstate competition to host segments of (hyper)mobile capital. States act to adjust their conditions (laws, etc.) to retain or create comparative advantage in attracting certain segments of capital. It is possible that a state would offer its GIS to capital choosing to locate there (e.g., for use in marketing, location, or labor relations decisions), thus providing an incentive in much the same way as a tax holiday.

In a wider sense, the depiction of the globe is often used to show the home of humankind as a singular, limited, fragile entity—and as such it is often seen in environmentalist materials. However, in the advertisements under consideration here, the world is not a cozy unity.

Rather it is a manipulatable world shattered into bits of information, albeit holding the promise of reconstitution. This world of information potentially contains everyone and everything—with or without their assent. The masculine will entailed in the visual capture and representation of earth space results in a technology-enabled representation of space that is available only to a few viewers or users. The vast majority of the earth's inhabitants do not know their homes are for sale, to be purchased and represented on computer screens, as spatial data (geographic information). It is without irony that one commentator has noted the possibility for a role to be played by Third World nations:

> Thus there are many ways that less developed countries can play an extremely important part in data-collection efforts. Space assets operated by a few countries plus ground-truth supplied by many others can give us the worldwide picture we need. In some cases, the equipment to collect the data may be supplied at no cost to the host countries by the spacefaring nations which need the resulting data to round out their own information on worldwide environmental concerns. (McLucas, 1991, p. 138)

The questions, Information for whom? and To what end? are either not asked or are assumed unproblematically. The complex and contested inequities in the human world of social *spatial practices* are made more orderly in the *representational spaces* of GIS and in GIS themselves as *representations of space.*

CONCLUSION

In the case presented here, Lefebvre's tripartite scheme provides a rigorous but flexible framework for assessing GIS and associated technologies as ways of knowing, seeing, ordering, and reproducing the material lived world and social relations. This chapter began with advertisements for geographic technologies as *representational spaces.* Advertisements may be the most "public" face of geographic technologies. These *representational spaces* are more than decorative, clever, compact, visual images. The advertisements are neither innocent nor incidental: They embody and reproduce the assumptions and claims of geographic technologies. In our analysis, magazine advertisements for geographic technologies are just one moment in the ongoing social dialectic surrounding GIS and associated technologies. The text and imagery in the advertisements can be scrutinized as claims that, as they structure the *representational spaces,* they also are implicated in linking the constitution of the *representations of space* (GIS, etc.) and *material spatial practices.* The print advertisements are thus seen as embedded in their context. Salient themes in the advertisements—power, control, manipulability, break up and recombination of data—are traced through in the *representations of space* themselves. GIS and associated technologies as *representations of space* do not matter only in terms of how they are involved in *representational spaces,* but are part and parcel of contemporary *spatial practices.* This link can be critically investigated in terms of the political economy of GIS and in their applications—the cold war, gendered space technology, multinational corporate strategies—many of which are celebrated in the technologies' *representational spaces.*

Geographers have a long-standing interest (not always critical) in the representation of space. Presently, this interest is reemerging in a more critical guise at the same time as new, technology-based ways of representing the world (GIS, etc.) are being developed, employed, and increasingly becoming taken-for-granted *representations of space.* This chapter attempts to bring insights concerning the social production of space to bear upon GIS and associated technologies. Through a focus on one moment in their relational circuits, this chapter situates geographic technologies and claims made on their behalf within social relations—or more precisely, within the social production of space through *spatial practices.*

ACKNOWLEDGMENTS

We wish to extend our thanks to Heidi Nast, Pat McHaffie, and Tom Klak for their helpful comments.

REFERENCES

Aerial Data Reduction Associates, Inc. (1990). [Advertisement]. *ACSM Bulletin, 128,* 36.

Alvares, C. (1992). *Science, development and violence: The revolt against modernity.* Delhi, India: Oxford University Press.

American Digital Cartography. (1991). [Advertisement]. *Geo Info Systems, 1*(6), 9.

Autodesk, Inc. (1992). [Advertisement]. *Geo Info Systems, 2*(10), 23.

Barnet, R. J., & Müller, R. E. (1974). *Global reach: The power of the multinational corporations.* New York: Simon and Schuster.
Barry, J. (1992). Mappings—A chronology of remote sensing. In J. Crary & S. Kwinter (Eds.), *Incorporations: Zone 6* (pp. 570–571). New York: Urzone.
Barthes, R. (1979). *The Eiffel Tower and other mythologies* (R. Howard, Trans.). New York: Hill and Wang.
Bayley, S. (1986). *Sex, drink and fast cars: The creation and consumption of images.* London: Faber and Faber.
Benjamin, W. (1968). The work of art in the age of mechanical reproduction. In H. Arendt (Ed.), *Illuminations* (pp. 217–252). New York: Schocken Books. (Original work published 1936)
Benjamin, W. (1980). A short history of photography. In A. Trachtenberg (Ed.), *Classic essays on photography* (pp. 199–216). New Haven, CT: Leete's Island Books. (Original work published 1931)
Berger, J. (1972). *Ways of seeing.* London: British Broadcasting Corporation/Harmondsworth, U.K.: Penguin Books.
Canadian Marconi Co. (1992). [Advertisement]. *GPS World, 3*(8), 59.
Cosgrove, D. (1988). Prospect, perspective, and the evolution of the landscape idea. *Transactions of the Institute of British Geographers* (New Series), *10,* 45–62.
Cosgrove, D. (1989). Looking in on our world: Images of global geography. In P. Wombell (Ed.), *The globe: Representing the world* (pp. 13–18). York, U.K.: Impressions.
Crary, J. (1990). *Techniques of the observer: On vision and modernity in the nineteenth century.* Cambridge, MA: MIT Press.
de Certeau, M. (1984). *The practice of everyday life* (S. Rendall, Trans.). Berkeley and Los Angeles: University of California Press.
Digital Equipment Corp. (1992, July). [Advertisement]. *GIS World* [Special Issue], *67.*
Digital Matrix Services. (1991). [Advertisement]. *Geo Info Systems, 1*(7), 23.
Driver, F. (1992). Geography's empire: Histories of geographical knowledge. *Society and Space, 10,* 23–40.
EDS. (1992). [Advertisement]. *Geo Info Systems, 2*(7), 17.
EOSAT. (1991). [Advertisement]. *Geo Info Systems, 1*(6), 21.
Epson America Inc. (1992, December). [Advertisement]. *PC World,* p. 57.
Erdas. (1991). [Advertisement]. *Geo Info Systems, 1*(3), 7.
Flink, J. J. (1975). *The car culture.* Cambridge, MA: MIT Press.
Foucault, M. (1977). *Discipline and punish* (A. Sheridan, Trans.). London: Allen Lane.
Gardner, N. (1992). Remote sensing. In A. Rogers, H. Viles, & A. Goudie (Eds.), *The student's companion to geography* (pp. 151–159). Oxford: Basil Blackwell.
Genasys. (1990). [Advertisement]. *GIS World, 3*(6), 16.
Giddens, A. (1985). *A contemporary critique of historical materialism, Vol. 2. The nation-state and violence.* Berkeley and Los Angeles: University of California Press.
Haraway, D. (1991). *Simians, cyborgs, and women: The reinvention of nature.* New York: Routledge.

Harvey, D. (1989). *The condition of postmodernity.* Oxford: Basil Blackwell.
Hepworth, M. (1990). *Geography of the information economy.* New York: Guilford Press.
Horizons Technology, Inc. (1992, December). [Advertisement]. *PC World,* p. 305.
Intergraph. (1992). [Advertisement]. *Geo Info Systems, 2*(3), 17.
Keller, E. F. (1992). *Secrets of life, secrets of death: Essays on language, gender and science.* New York: Routledge.
Lefebvre, H. (1991). *The production of space.* (D. Nicholson-Smith, Trans.). Oxford: Basil Blackwell. (Original work published 1974)
Livingstone, D. N. (1992). *The geographical tradition: Episodes in the history of a contested enterprise.* Oxford: Basil Blackwell.
Mack, P. E. (1990). *Viewing the earth: The social construction of the Landsat satellite system.* Cambridge, MA: MIT Press.
Massey, D. (1992). Politics and space/time. *New Left Review, 196,* 65–84.
McLucas, J. L. (1991). *Space commerce.* Cambridge, MA: Harvard University Press.
Meinig, D. W. (1979). Symbolic landscapes: Models of American community. In D. W. Meinig (Ed.), *The interpretation of ordinary landscapes* (pp. 164–192). New York: Oxford University Press.
Mitchell, W. J. T. (1986). *Iconology: Image, text, ideology.* Chicago: University of Chicago Press.
Nandy, A. (Ed.). (1988). *Science, hegemony and violence: A requiem for modernity.* Delhi, India: Oxford University Press for United Nations University, Tokyo.
PCI Remote Sensing Corp. (1992). [Advertisement]. *Geo Info Systems, 2*(10), 25.
Qualitas. (1992, December). [Advertisement]. *PC World,* p. 67.
Radian Corp. (1992). [Advertisement]. *Geo Info Systems, 2*(7), 23.
Rae, J. B. (1971). *The road and the car in American life.* Cambridge, MA: MIT Press.
Ramtek. (1991, November/December). *Geo Info Systems,* p. 17.
Raper, J. (1992). Geographical information systems. In A. Rogers, H. Viles, & A. Goudie (Eds.), *The student's companion to geography* (pp. 168–178). Oxford: Basil Blackwell.
Schulman, R. D. (1991). Portable GIS: From the sands of Desert Storm to the forests of California. *Geo Info Systems, 1*(8), 25–33.
Soja, E. (1990). *Postmodern geographies: The reassertion of space in critical social theory.* London: Verso.
SPOT Image Corporation. (1992). [Advertisement]. *Geo Info Systems, 2*(5), 9.
Strategic Mapping, Inc. (1991). [Advertisement]. *Geo Info Systems, 1*(7), 11.
Thrift, N. (1990). Commentary: The perils of the international financial system. *Environment and Planning A, 22,* 1135–1140.
Trachtenberg, A. (1991). Photography: The emergence of a keyword. In M. A. Sandweiss (Ed.). *Photography in nineteenth-century America* (pp. 17–47). Fort Worth, TX: Amon Carter Museum/New York: Harry N. Abrams.
Trimble. (1990, November–December). [Advertisement]. *Geo Info Systems, 1,* 14.

CHAPTER 9

Pursuing Social Goals Through Participatory Geographic Information Systems

REDRESSING SOUTH AFRICA'S HISTORICAL POLITICAL ECOLOGY

Trevor M. Harris
Daniel Weiner
Timothy A. Warner
Richard Levin

There is growing debate concerning the concept and practice of development (Sachs, 1992; Watts, 1993). Traditional developmentalism is being criticized for being a Western product that perpetuates social and spatial inequality because it is market-driven, technology-based, resource-intensive, and undemocratic. Poor people and places, it is claimed, are "transmogrified into an inverted mirror of others' reality" (Esteva, 1992, p. 6). This attack on the discourse of developmentalism is relevant to the emerging debate regarding the role of geographic information systems (GIS) in society. GIS utilization for research, planning, and project assessment has generally come to be seen as a technicist legitimation of the historical power relations associated with traditional developmentalism.

In this chapter, GIS and remote sensing technologies are employed to explore the intersection of political ecology and participatory policy formulation in the postapartheid reconstruction of South Africa. Because apartheid was in essence a geographical project, GIS could be very useful in helping to redress South Africa's spatial organization of production and its currently highly skewed distribution of resources. The chapter seeks to contribute to the discourse on developmentalism and to the dialogue about the use of GIS in human geography through a case study of alternative GIS and remote sensing applications in a rural locality within the Eastern Transvaal region of South Africa. We report on the very early phases of a participatory research and policy formulation effort and raise a number of issues related to the strengths, weaknesses, and potential contradictions associated with "progressive" community-based GIS utilization.

Through the Eastern Transvaal case study, we argue for a participatory process of social transformation which employs advanced digital technology. Our argument for a participatory GIS is intended to demonstrate a GIS application where local knowledge, community needs, and specific social histories are appreciated and incorporated into the development process, and "expertise" is viewed as interactive. In this way the production of information is viewed not solely as a top-down operation but one whereby local knowledge arising from social narratives is converted into data within a GIS for research and policy formulation. The chapter also contributes to the growing debate within geography regarding the articulation of GIS, power, and knowledge.

DEVELOPMENTALISM DISCOURSE

For more than a century, the idea of development symbolized social progress. Capitalism, it was generally agreed, was central to this process. According to Esteva (1992, p. 9), "Both the Hegelian concept of history and the Darwinian concept of evolution were interwoven in development, reinforced with the scientific aura of Marx." For some Marxists the success of the socialist project necessitated an initial capitalist stage. As a result of what Slater (1992, p. 307) calls "ethnocentric universalism,"

> Development has . . . rarely broken free from organicist notions of growth and from a close affinity with teleological views of history, science and progress in the west. . . . By the nineteenth century the central thesis of developmentalism as a linear theory of progress root-

> ed in capitalist hegemony was cast in stone; it became possible to talk of societies being in a state of "frozen development." Alternatives to classical development thinking—dependency, Marxisms of various sorts—frequently shared the economism, linearity, and scientism of "developmentalism." Their universalism carried the appeal of secular utopias constructed with the bricks and mortar of rationalization and Enlightenment (Watts, 1993, p. 259).

GIS is often alleged to be both a product of this historical discourse and a source of its late-20th-century social reproduction. Implicit in the latter claim is the assumption that GIS reifies Western definitions of knowledge and meaning represented as technical data in a computer system which represents the sum of "expert" knowledge.

The continual production and reproduction of global poverty, combined with the gradual peripheralization of some "core" regions, the collapse of "existing socialism," and the varied influences of postmodernity has thrust developmentalism into conceptual and epistemological confusion (Sachs, 1992). From radical discussions about development as destruction,[1] to neoliberal concerns about the "balance between state, market, and civil organizations" (de Janvry, Sadoulet, & Thorbecke, 1993, p. 565), the discourse of developmentalism itself is in transition. Watts (1993, p. 258) suggests that the new "antidevelopmentalism proposes not so much a development alternative as an alternative to development." The potential for New Social Movements (NSM) to forge "new social contracts between state and associational life . . . has led some to suggest that in the belly of the world capitalist beast is a 'new mode' of doing politics, new sorts of fragmented subjectivities, and a bottom-up horizontal vectoring which redefines political and economic democracy" (Watts, 1993, p. 268). For Escobar (1992, p. 412), the Third World must "shake off the meanings imposed on them . . . to open up in a more explicit manner the possibility for a different regime of truth and perception within which a new practice of concern and action would be possible." Antidevelopmentalism is, therefore, a call for an "actor-oriented paradigm" of social change (Long, 1992, p. 20) where decentralized political activity provides an alternative to the antiquated developmentalism of modernity.

A number of issues can be raised as to whether the concepts and practices of grassroots/community development and participation, sustainability, and local empowerment actually represent an alternative to development, or merely represent a new discourse within an old developmentalism. Are NSM really that new? Is the recent academic preoccupation with NSM part of the "discovery" of the local? And, what about the "relations between NSM and the hegemonic class forces of capitalism" (Watts, 1993, p. 268) as powerful influences from the Right (e.g., the World Bank; see Mackenzie, 1992) that are also concerned with the relationship between civil society, the state, greater grassroots participation, and the localization of the development process? Antidevelopmentalism is a struggle over the definitions and practices of development rather than a thorough rejection of the modern project itself. For the most part, NSM are not struggling against developmentalism, per se, but for localized development definitions and practices and transformed configurations of power and access associated with "different modernisms" (Pred & Watts, 1992, p. xiv). Roads, water and sanitation systems, electricity, land, and jobs remain central components of NSM. What is "new" is our (e.g., academics, development consultants, planners) terrain of discourse and a growing appreciation of everyday life and the politics of scale. As Pred (1992) notes:

> The intersection of economic restructurings—of historically and geographically specific forms of capitalism—with already existing localized patterns of everyday life almost inevitably elicits some form of cultural response or negotiation, some overt or covert form of symbolic contestation. The repeated but varied cultural responses that emerge as part of local transformation, as part of the local introduction (or removal) of specific forms of capital . . . [are] . . . part of the articulation of the local and the global. (p. 107)

It is this articulation of scale, and the powers associated with differing forms of knowledge, expertise, and meaning, that lie at the heart of our discussion of participatory GIS and postapartheid transformation in South Africa. Rural blacks desire (and demand) to participate in the policy formulation process as a "cultural response" to generations of colonial and apartheid repression and social engineering. In South Africa, an overly technicist and top-down policy environment, supported by GIS, is likely to reinforce current neoapartheid forms of rural restructuring. Conventional developmentalism will also be very unpopular in poor rural communities (Levin & Weiner, 1993). In pursuing a participatory approach to the use of GIS and a focus on the inclusion of social narratives and behavioral information, our work is an attempt to utilize GIS in the process of bottom-up social transformation. This path, as will be demonstrated, is littered with numerous difficulties and potential contradictions.

GIS, DEVELOPMENTALISM, AND HUMAN GEOGRAPHY

The public debate about the role of GIS in geography recently initiated by Taylor (1990), Openshaw (1991), Goodchild (1991), Taylor and Overton (1991), Pickles (1991), and Edney (1991) has shifted the early focus on GIS away from a predominantly technical or applications base toward a dialogue about the use of GIS and its relations to ongoing theoretical advances in human geography. In some respects this dialogue has antecedence in the GIS community with the concern for the organizational framework of operational GIS and with the potential legal issues associated with the use or misuse of GIS data and products. However, the sophistication and range of the debate now extends far beyond the boundaries of those earlier discussions. The clarion call of the present debate within geography is that GIS be evaluated "warts and all." Given the lack of critical evaluation of GIS by other disciplines, this debate is also important for disciplines beyond the borders of our own. Following the initial polarization of the debate, discussion is now progressing toward a more balanced assessment of the issues raised by these early exchanges. We draw upon the tensions arising from the current GIS/geography/society debate to address some of these issues in the context of alternative policymaking initiatives in South Africa.

In the application of GIS to our work in South Africa four main themes are developed. The first theme concerns issues of equity and equality. Chrisman (1987, p. 1367) presaged this aspect of GIS when he observed that "as . . . our technology finds its way into practical use, it must be accountable economically . . . politically, socially, and even ethically." His call for the design of GIS based on social and cultural goals was rooted in a concern (prophesy) that "some of the current successes [of GIS] may have been achieved by exploiting the easy parts of the problem. There is a danger that the tough issues, temporarily swept under the rug, will reemerge, perhaps to discredit the whole process" (1987, p. 1367). A dominant theme of our work in postapartheid South Africa is clearly related to the utilization of GIS in the pursuit of equity. As Chrisman (1987) argues,

> Geographic Information Systems should be developed on the primary principle that they will ensure a fairer treatment of all those affected by the use of the information (equity). Certain solutions, though efficient in their use of computing, do not support the effective use of institutions or the equitable results of the analysis. (p. 1367)

Quite how equity, rationality, and objectivity are handled in a GIS, however, is problematic (see Openshaw, 1991; Taylor & Overton, 1991). The pursuit of equity demands that political decisions be made as to what constitutes fair treatment. At the macrolevel and in the context of postapartheid South Africa, the issue of equity is clearly fundamental to any use of GIS to generate and support policy initiatives. The pursuit of social goals through GIS is a political process and cannot ignore this fact, however much GIS may allow us to simulate possible alternative decision-making scenarios. A GIS reflects the mandate of the agency that operates it. Agencies have internal rules and value systems, as well as a stake in self-preservation. The extent to which GIS represents objectivity in terms of what data is used, or how it is classified, or how it is analyzed, or the interpretations drawn from it, is clearly highly questionable. Value-neutral GIS simply do not exist. GIS must be viewed as an extension of institutional goals; this truth is crucial to understanding the use and operation of a GIS in any development project. GIS operationalizes the mandate of an agency and is an integral instrument in defining and implementing agency decisions with regard to equity. As Chrisman (1987) counsels, the effectiveness of a GIS should be evaluated within the context of socially driven goals that link the dual principles of equality and equity.

GIS is also integrally linked to issues of equality in terms of access to data, information, and knowledge. This is especially pertinent to South Africa, where there is a profound apartheid legacy in information generation and dissemination. The focus by Taylor (1991) and Taylor and Overton (1991) on distortion and bias in knowledge production is especially relevant to the linking of GIS to traditional developmentalism in the world's periphery. Their arguments that structural knowledge distortion exists because of the selective participation of groups in data/information production, and that this distortion reflects the dominance of the Western world in global power relations and the exclusion of nonhegemonic forms of knowledge, must be admitted to be true. As Taylor (1991, p. 89) notes, "Small groups such as indigenous peoples have only recently had their voice heard in a social science that has been wedded to 'modernization.' As such, our social science has appeared more and more like a story of 'winners' and cultural imperialism." In the context of "traditional" developmentalism, GIS can be interpreted as representing a continuation and reinforcement of this structural distortion of knowledge. In the mode of top-down data creation, GIS empowers the powerful and disenfranchises the weak and not so powerful via the selective participation of groups and individuals.

A number of crucial issues thus arise as to how local empowerment could proceed. Furthermore, how can local knowledge be represented to redress the structural knowledge distortion of traditional GIS-based developmentalism? Such questions pose substantive problems for those seeking to utilize the strengths of GIS technology in the pursuit of social goals, particularly in the context of differentiated community power relations. These issues revolve not only around the functional limitations of the technology itself and its inability to handle knowledge, as opposed to data, but also around the dissemination and use of a technology by groups without GIS expertise. GIS becomes, in itself, part of the debate about developmentalism and is integral to issues of local democracy and empowerment.

Pickles (1991), quoting Illich (1985), provides a powerful metaphor about the impact on social relations resulting from the introduction of a loudspeaker (technology) onto a Greek island that is very useful in this context. Whereas previously on the island, access to the public and public speech had been an open resource, the introduction of a new technology effectively privatized public speech and transformed the public space of discussion into a hierarchical space delimited by access to the loudspeaker. As Pickles (1991, p. 84) remarks, "For users of the technology, access to information and other users is greatly enhanced. But adoption also implies non-adoption or inability to adopt. Polarization of users and non-users results. Any assertion of the democratic nature and use of GIS must address this issue." In the light of the power relations associated with data access and knowledge distortion, the process of transforming local social, environmental, and political knowledge into GIS data is an important challenge to the GIS community.

A third important theme is related to differential access to GIS data, technology, and expertise. The establishment of a GIS database and the acquisition of hardware, software, and trained personnel is an expensive process. These costs usually limit GIS technology to state agencies or large private corporations. The conditions, or preconditions, that regulate access to that information also usually reside with the same agencies. With the continued diffusion of GIS into development planning, the issue of unequal access to data, technology and expertise is likely to reinforce the political and economic status quo and to work against more equitable planning decisions. As a recent report on the use of GIS in international development comments, "It is impossible to have sustainable and equitable development without free access to reliable and accurate information" (Benmouffok, 1993, p. 4). Without equitable access to GIS data and the technology, small users, local governments, nonprofit community agencies, and nonmainstream groups are significantly disadvantaged in their capacity to engage in the decision-making process (Edney, 1991). Independent of the resource and expertise base needed to utilize GIS data, there are also pertinent power-resource and access implications between the various actors associated with GIS implementation and the recipients of development. Not least in the context of top-down GIS planning, the "discovery" of local resources contextually (Yapa, 1991; Edney, 1991) and the production of alternative local knowledge is greatly constrained. If careful attention is not paid to these issues, the resulting GIS will necessarily mirror the agency/funder mindset and value system. The incorporation of local knowledge, however, may or may not support alternative options which counter exploitation by governments and corporations (Edney, 1991, p. 101). Quite clearly, access to knowledge production has important ramifications for the empowerment or disempowerment of specific social groups within communities. The free flow of information is essential to truly democratic implementation of GIS.

Assuming that the free flow of information is unlikely, Pickles (1991) raises a fourth issue regarding the surveillant capabilities of GIS by the academy, state, and capital, and the potential impact of the technology on restructuring global, regional, and local geographies. In linking GIS with the interests of capital, he challenges claims that GIS fosters democratic practice, or broadens the distribution of, or access to, information. Pickles suggests that the term "information society" is a misnomer that hides the increasing surveillant capability of state institutions and transnational corporate enterprises. He identifies speed of access, privacy, ownership, and control of social action as representing areas that run counter to efforts at broadening local community accessibility and democratic control. Pickles's argument implies that the increasing demands for more data by communities, even with open access privileges, may indirectly support the development of a surveillant technology due to the power relations associated with GIS data use.

The issues of GIS and equity, differential access to data and technology, and concerns about a surveillant capability are especially manifest and exacerbated in the context of South Africa. Grand apartheid was a geographical project that consolidated colonial processes of uneven development at local, regional, and national scales. A redistributive postapartheid policy environment will, therefore, necessitate that national issues of social justice be connected spatially and historically to specific social histories and political ecologies. National and regional policies and programs must integrate local needs and capacities.

THE POLITICS OF REDRESSING SOUTH AFRICA'S HISTORICAL POLITICAL ECOLOGY

The historical implementation of apartheid was based on the production and utilization of spatial data. The forced removal of blacks from designated "white" territories and their relocation into urban townships and rural bantustans was planned, organized, and researched through the institutions of the apartheid state. In 1950 one-third of the total African population lived in "white" rural areas. By 1992 the proportion had declined to 11% (South African Institute of Race Relations, 1993). The Surplus People Project estimated that 3.5 million removals took place between 1960 and 1983 (Platzky & Walker, 1985); remarkably, this is an underestimation of total forced removals. Apartheid-forced removals solidified a much longer colonial process of (black) land dispossession, the development of (white) capitalist agriculture, and a space-economy with abundant supplies of cheap migrant labor. South Africa's contemporary poverty and socioenvironmental geography must be viewed within this historical context.

Presently, the bantustans account for approximately 17% of South Africa's total area and yet accommodate 57% of the African and 44% of the total population (Pickles & Weiner, 1991). Access to good quality land in these "homelands" is extremely limited, and in many locations there is a severe water, fuelwood, and biomass crisis. One-half of South Africa's total population of 40 million and nearly two-thirds of the African population of 30 million, can be classified as living "below the poverty datum line'' (South African Institute of Race Relations, 1993, p. 198) or ''below subsistence'' (Wilson & Ramphele, 1989, p. 17). Approximately 7 million people (mostly Africans) presently live in shacks (Smith, 1992), while more than three-quarters of rural Africans are classified as poor. The majority of rural blacks[2] are landless or near-landless (Pickles & Weiner, 1991). Redressing South Africa's historical political ecology must be a central component to any attempt at reducing postapartheid poverty and uneven development. South Africa's black population needs additional land for residential, agricultural, and grazing purposes, and water is in short supply both for production and consumption. Very high levels of unemployment and stagnant formal job creation nationally accentuate the need for rural resources for rural household reproduction. The history of forced removals provides an emotional, moral, and political urgency to the materiality associated with contemporary rural social and spatial marginalization.

There is a growing debate on how postapartheid rural reconstruction should take place. In transitional South Africa, the various social forces which constitute the new state recognize the importance of access to and control over spatial information for postapartheid planning. In redefining and planning the future landscape, spatial data and information represents a valuable commodity. All the major political parties in South Africa now recognize the potential applicability of GIS for electoral purposes, regional planning, and land reform. Access to GIS and data, therefore, has become highly politicized. The privatization of many data-collecting and data-providing state agencies is doubtless related to this. The African National Congress (ANC), which is making the difficult transition from liberation movement to the leading political party in the government of national unity, is at a disadvantage, not least in its access to expertise, technology, and existing data. In this ''new'' South Africa, potential GIS applications are being conceptualized within the context of traditional developmentalism.

There are four main points regarding South Africa's rural land reform debate that are particularly relevant to this chapter. First, with the exception of the Far Right, there is a consensus that rural land reform is necessary. The Development Bank of Southern Africa (DBSA), an influential state development funding institution, argues that ''a land reform program, encompassing social, political, economic, technical and environmental aspects, and consisting of an entitlement and empowerment approach, is both desirable and feasible'' (Brand,

Christodoulou, van Rooyen, & Vink, 1992, p. 374). The ANC also calls for the implementation of a widespread land redistribution program where "all South Africans are entitled to equitable access to land and shelter" (African National Congress, 1992, p. 27). The World Bank, which is moving rapidly into the postapartheid policy arena, argues for "a Rural Restructuring Program (RRP) that would achieve a rapid transfer of a significant share of agricultural land to black users" (International Bank for Reconstruction Development, 1993, p. 1).

Quite how this reform process will be implemented is presently not known. Certainly, it would appear that a top-down process is envisaged even though the proposed DBSA and World Bank mechanisms for land transfer differ significantly from the ANC's public and democratically derived position. The World Bank argues, for example, for "a design that relies as much as possible on an open and accessible land market, but with a strong affirmative action component in the form of broadly targeted injections of purchasing power." The DBSA (Brand et al., 1992, p. 372) concurs: "The land market, supplemented by affirmative action programs, will have to be the major instrument for structural reform in a strategy for providing for full access to land." This suggests that there is a real danger that GIS adoption for rural land reform in South Africa will legitimize policies emanating from historically conservative development institutions.

Second, there is mounting evidence that important elements within the ANC leadership are also moving closer to the World Bank/DBSA position regarding appropriate mechanisms for land transfer. Compromises associated with political negotiations, combined with a growing petit-bourgeois orientation within the hierarchy of the organization, are solidifying a technicist and neoclassicist policy orientation (see Levin & Weiner, 1993). In July 1993, two of the authors of this chapter were told by an influential ANC official working on rural restructuring policies that "we will have to accept the reality of market-based land reform." This is an important first step in the familiar process whereby internal African policy agendas are strongly influenced by international institutions within the context of a policy alliance between domestic and international capital.

Third, market-based land reform, combined with a clause protecting existing private property that will be part of the new constitution, will protect the existing agrarian bourgeoisie and privileged wealthy blacks as new entrants into farming. It is also likely to close off options for acquiring high-potential land while generally poor-quality land is made available for resettlement (this was the case in Zimbabwe; see Weiner, 1991). Market-based land reform will do little to help the majority of victims of forced removals and will protect the rural beneficiaries of grand apartheid. In the short-to-medium term, the geography of a World Bank/DBSA/ANC policy alliance with respect to private property protection and market-restricted access to land will likely contribute to the reproduction of South Africa's historical political ecology. This is precisely why the National Party (NP), which was in power since 1948, and the ANC are striving for operational GIS capability. The major political parties understand that information is power.

Fourth, regarding rural land reform, tension between the ANC hierarchy and its constituency could become a problem. The demobilization of an historically mass-based organization is taking place as the policy debate is transformed by the politics of elite bargaining. As a result, the process of reconstruction will prove to be difficult as neo-apartheid geographies (see Pickles & Woods, 1992; Weiner & Naidoo, 1993) become entrenched within a top-down and market-driven policy environment. Key land reform decisions are being made by South African and international "experts" with very little incorporation of local community knowledge and expertise. The victims of apartheid have been essentially excluded from the debate. As a result, the parameters of future policy and definitions associated with the emerging rural land reform discourse are being shaped in a way that will make successful local implementation difficult to achieve. For example, on the issue of who should be eligible for rural land reform, there has been a tendency to target only the victims of forced removal while inaccurately assuming generally poor and homogeneous rural communities. Our participatory research indicates significant social differentiation within rural communities, with a demand for land reform which goes well beyond households recently forcibly removed. While development "experts" (somewhat frantically) search Africa and the globe for appropriate land reform "models," rural communities have been articulating rather sophisticated views of future land use.

This legitimation of traditional developmentalism raises a series of interesting questions and concerns regarding the utilization of GIS for rural land reform in South Africa. For example, how can this advanced technology be incorporated into a community-based participatory planning process where local knowledge is incorporated and analyzed interactively? The potential even for participatory GIS to legitimize an essentially top-down technicism is an additional concern; participation generally is being propagated precisely for this purpose. The line between GIS as a potentially liberating policy formulation framework and a technology that serves to reproduce existing power relations can be very unclear. These key issues are addressed in a case study of the Kiepersol area within the Eastern Transvaal region.

THE KIEPERSOL CASE STUDY

Kiepersol is one site within a larger participatory research and policy formulation program. Work to date has pointed to the potential benefits of employing advanced digital spatial data handling technologies for rural land reform in South Africa and led to the proposed development of a participatory GIS. We reiterate again that the work being presented here is preliminary and focuses on the major technical, political, and practical aspects of establishing a participatory GIS for the Kiepersol area. The case study also highlights a number of potential applications and contradictions.

The Kiepersol locality, in the Lowveld of the Eastern Transvaal, has significant social and ecological variation and a long history of contested resources and forced removals (Mabin, 1991). Between 1960 and 1982 at least 381,000 people were relocated to KaNgwane, mostly from farms in "white" areas of the Transvaal (Platzky & Walker, 1985). A major cause of this relocation was the abolition of labor tenancy, which resulted in worker/tenants suddenly becoming illegal squatters on the land on which their families had lived for many years. As recently as the period between 1985 and 1991, the number of people living in the Nsikazi district of KaNgwane increased by an estimated 5.3% per annum. In 1991, population density was 356 people per square kilometer compared to an average density of 35 in the nearby "white" Nelspruit Magisterial District (DBSA, unpublished data). As a result, rapid bantustan peri-urban development has occurred between the unpopulated Kruger Park in the east, and the sparsely populated white farming area to the west, which is a portion of the White River district (subsequently referred to as White River). This is shown in Figures 9.1 and 9.2, which depict the local manifestation of grand apartheid geographies in the uneven distribution of land, settlement, major roads, perennial rivers, and dams.

The stark contrasts associated with apartheid land distribution and

PLATE 1. False-color-composite of May 3, 1992 SPOT data of Kiepersol, Eastern Transvaal. The image covers the same area as Figures 9.1 and 9.2. In this false-color rendition, green is displayed as blue, red as green, and near-infrared as red. (©SPOT Image. Copyright 1992 CNES.)

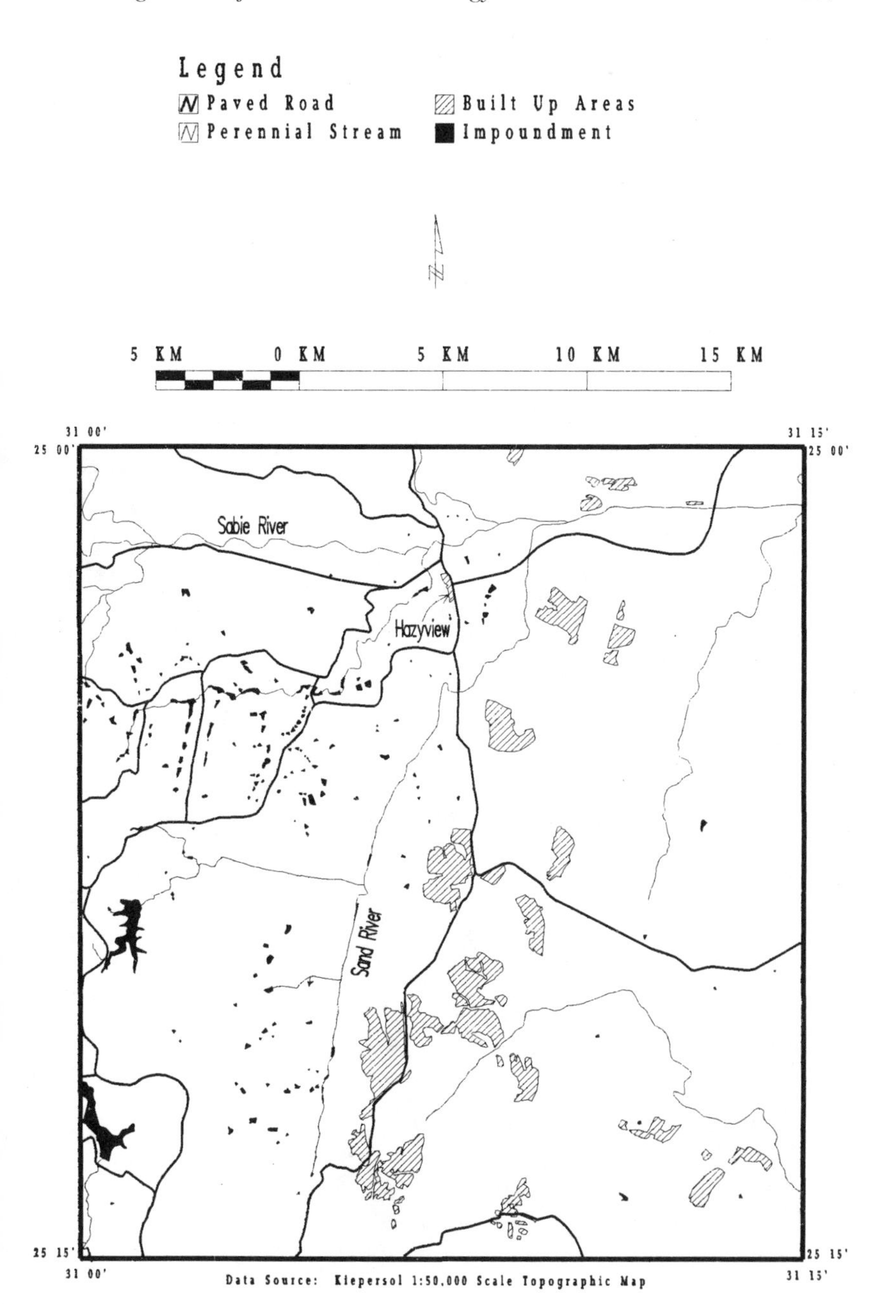

FIGURE 9.1. Map of hydrology, urbanization, and roads of Kiepersol, Eastern Transvaal, South Africa. From Chief Director of Surveys and Mapping (1984).

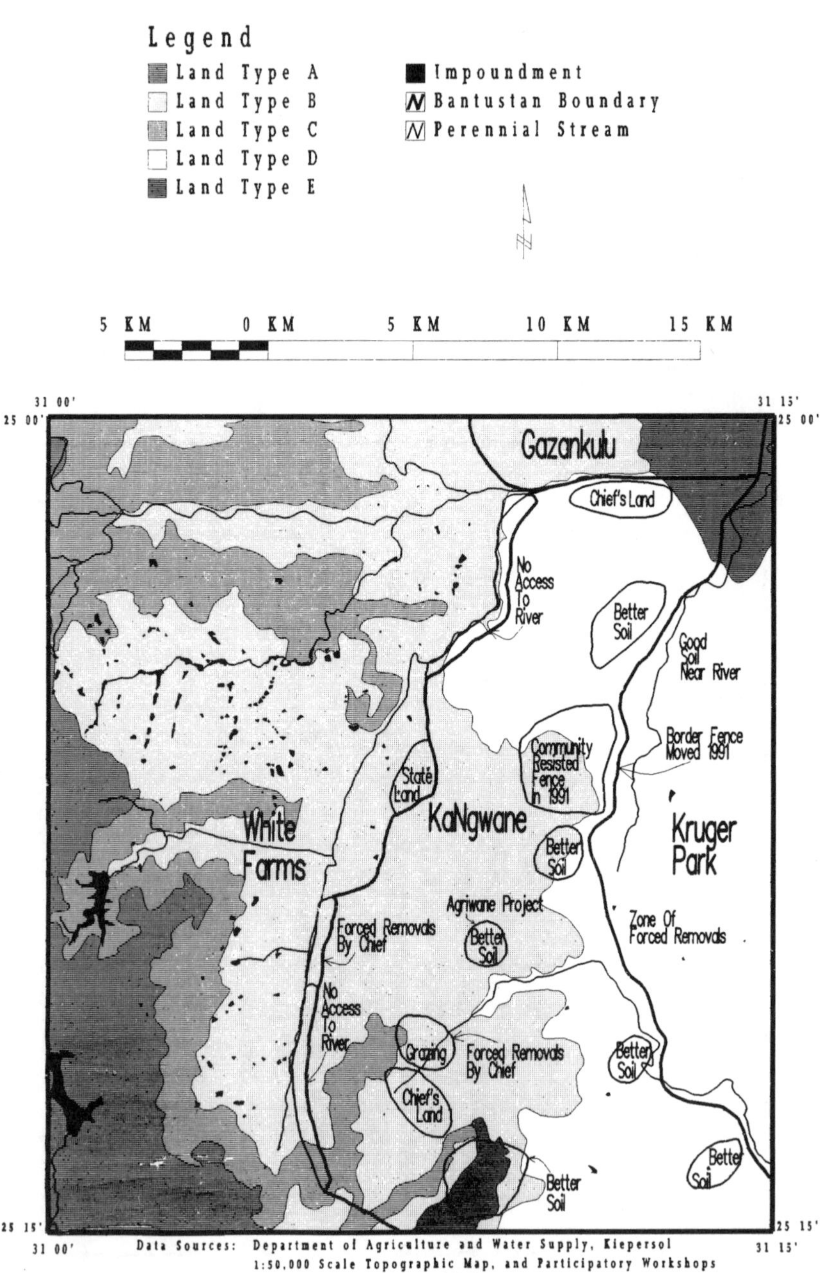

FIGURE 9.2. Map of Land Types and Mental Map of Kiepersol, Eastern Transvaal, South Africa. From Chief Director of Surveys and Mapping (1984). Land types summarized from Land Type Survey Staff (1989); mental map from participatory workshops.

use are also illustrated in the SPOT satellite image shown in Plate 1. In this false-color image, strong near-infrared reflection, which is characteristic of verdant vegetation, is shown in bright red. The variations in tones and colors are evidence of the steep precipitation gradient and marked differences in land potential, settlement type, and production relations within this relatively small area. Three distinct zones are apparent in Plate 1. The semiarid Kruger Park in the east has an average precipitation of less than 700 mm per annum and is associated with a comparatively dry, open savanna. The Nsikazi district of the KaNgwane bantustan located in the center of the image is not as dry. However, the denuded vegetation and dense rurban settlements result in higher spectral reflectance, and thus brighter and lighter tones on the image. Lush vegetation within Nsikazi is limited to the river valleys. The third and last zone is the White River region, which occupies the western portion of the area in the image. The climate is subhumid, with an average rainfall ranging from 860 mm in Hazyview to over 1,300 mm in the hills to the extreme west. The area is characterized by forestry plantations on higher elevations and irrigated banana, avocado and fruit production fields around the town of Hazyview. There are remnants of indigenous bush on the steeper slopes.

Access to water is an important component of contemporary politics and land use in Kiepersol and has been central to the historical geography of forced removals. White River is located upstream from KaNgwane on the major perennial river in this area, the Sabie, and has early access to the surface water resource of the area (see Figures 9.1 and 9.2). Furthermore, the eastern border of White River parallels a

---

Land Type A. Predominantly deep, well-drained sandy clay loams, with a medium to low base status, and a significant slope constraint for arable cultivation. Average rainfall 800 to 1300 mm per year.
Land Type B. Predominantly deep, well-drained sandy clay loams, with a medium to low base status, and only minor slope constraint for arable cultivation. Average rainfall 800 to 1300 mm per year.
Land Type C. Predominantly well-drained sandy clay loams, with medium to low base status, and a significant cultivation constraint due to rocky or shallow soils. Average rainfall 1000 mm per year.
Land Type D. Predominantly deep, well-drained coarse sands to loamy sands, with a high base status, but low clay content and thus a limited capacity for nutrient and water retention for dryland cultivation. Average rainfall 700 mm per year.
Land Type E. Soils predominantly too shallow and stony for cultivation. Average rainfall 700 mm per year.

drainage divide that limits flow from the higher rainfall escarpment region to the central Nsikazi area. Whereas agriculture in White River is typically capital-intensive, mechanized, and organized around dam-based irrigation, the bantustans are dominated by extensive dryland agriculture and livestock husbandry. Only a small minority of black farmers have access to irrigation, fertilizers, or tractors. Most residents of KaNgwane do not have direct access to water for domestic use in their homes. Most food is imported and nutrition is poor. The bantustan is, however, an important source of casual labor for White River farms and plantations, and also a source of cyclical migrant labor.

The satellite image provides a graphic illustration of the current distribution of resources in Kiepersol. But the quantification and mapping of resources, and the distribution of those resources, is highly political in transitional South Africa. The rural land reform debate is being conducted with very limited empirical information on the natural resource base and materiality associated with specific farming systems and forms of social organization at the local level. Broad, national-level generalizations, often with questionable validity at any scale, are being forwarded to support particular policy prescriptions. For some social groups, this information lacuna will be exacerbated by the privatization of government research and mapping agencies, which is resulting in rapid increases in the prices for data that was previously freely available from government agencies. Thus the transition to majority rule in South Africa may in fact result in even more limited public access to information.

Throughout South Africa there is a problem with maintaining data currency. This is due in part to rapid changes in the spatial organization of settlement resulting from the abolition of influx control legislation and rapid natural population growth. For example, extensive air photography collected in 1985 was inadequate to evaluate Kiepersol land use in 1993. This makes the use of satellite-based imagery vital. The digital nature of this type of data is an added advantage. The remote sensing of land use provides a powerful method of acquiring current information rapidly and of bypassing the state and its agencies. Yet beyond the already identified issues of cost, technology, and expertise, satellite imagery does not represent a panacea for the dearth of environmental data for the region. Land use activities, for example, must be inferred from the land cover. Furthermore, land use classes are by their very nature difficult to map, for they are generally small in area, spectrally variable, interspersed, heterogeneous, and exhibit seasonal and annual variation. It is often impossible to derive land use from land cover, for one cover type may have multiple possible land uses. Thus a land cover class of grassland may represent unused land, an abandoned field, a fallow field, grazing land, or energy from cattle dung. Many of these multiple uses might be occurring simultaneously and thus assigning a single land use class may be inappropriate. Despite these problems, it is nonetheless evident that great potential exists for integrating remote sensing applications into the participatory research and policy formulation process. Local knowledge is essential for interpreting the social production of space represented in a satellite image. We are in the early stages of such an effort.

Satellite-based remote sensing technology is also best suited to a broad perspective, but land reform is implemented locally. Matching the scale of data to the scale of question posed is a crucial step if ecological fallacy is to be avoided. Multiple interspersed land uses cannot be determined if land use is changing at a scale that is finer than the pixel size. Nor is increased spatial resolution always desirable since generally this means a trade-off with lower spectral and radiometric resolution. Remotely sensed images represent a temporal "snapshot" and seasonal variations affecting land cover and land use are complicated by the usual reliance on one scene. Cloud cover also tends to limit the coverages that can be obtained. The process of image classification, land cover mapping, and inferred land use require detailed knowledge of an area plus an understanding of seasonal and social patterns of land use. This again emphasizes the need for local knowledge, multidisciplinary expertise, and genuine community involvement for postapartheid reconstruction.

TOWARD THE DEVELOPMENT OF A PARTICIPATORY GIS TO SUPPORT GRASSROOTS INITIATIVES

In the 1990s, a range of ideological perspectives within the international development community are proclaiming the virtues of participation. For this reason we approach the concept and practice of participation with caution, fully conscious that it can be used in a context that serves to legitimize traditional developmentalism. In reality our study is concerned more with the development of a GIS for grassroots initiatives, of which participation is a primary objective, than with participatory GIS per se. The participation underlying this study is village-based and, as we stated, in a preliminary phase. After a large community-based meeting, a land committee of eight is elected and two of the eight are then chosen to represent the village at regional and national forums. Research assistants are hired from the community, and research questions, and methods are formulated and refined collectively in a series of local workshops. The process involves an interactive feedback format linked to the dissemination of research results, map construction, and discussion, as well as the articulation of a range of community policy issues and positions. Great care is taken in trying to account for differing community perspectives that may or may not be associated with class, gender, kinship, and age-based social differentiation. It is also important to note that the participatory process being initiated is, at present, being led from the top-down. Progress is being made, however, with local community "capacity building."

The primary objectives of the participatory GIS that is presently being established are: (1) enhanced community/development planner interaction in a research and policy agenda setting; (2) the integration of local knowledge with exogenous technical expertise; (3) the spatial representation of relevant aspects of local knowledge; (4) genuine community access to, and use of, advanced technology for rural land reform; and (5) the education of "expert" rural land use planners about the importance of popular participation in policy formulation and implementation. A participatory GIS approach seeks to develop a methodology where the social definition of land use potential is broadened to include the users of that space. A prescriptive, technicist program that ignores community perceptions and local knowledge by employing "textbook" definitions of agroecological potential and constraint is unlikely to address the true needs and capabilities of the inhabitants and may consequently be rejected or resisted. Our engagement in the process is not an attempt to "prove" that GIS can be used for progressive purposes. Rather, it reflects our own serious concerns regarding the power relations associated with GIS diffusion and the possibilities for alternative applications that support grassroots social transformation.

A central component of such a system is the incorporation of local knowledge as a thematic coverage. In KaNgwane village workshops, a mental map overlay was produced using the Kiepersol 1:50,000 topographic sheet and local village maps produced from air photography and field interviews (see Figure 9.2). Four groups of people (men and women with and without access to land) discussed the history of forced removals in Kiepersol, their perceptions of the quality of land, the location of "better soils," and the location of borders and other relevant aspects of land use and land potential in the Kiepersol area and their village. The mental map is a very partial representation of the many social processes and perceptions of space that exist. It does, however, display the many possibilities and problems associated with the spatial representation of local knowledge.

While the mental map is one way in which local knowledge can be captured and used in a GIS; it generates a number of problems that are as yet unresolved. First, the term "local knowledge" assumes a homogeneous community, subsuming complex local power relations and social differentiation into a single representation. In rural South Africa voices within communities are often filtered through the institutions of the chieftaincy and through lineage and gender status. The chiefs, who acted as structures of the apartheid state, remain powerful social actors in the countryside. They own a disproportionate amount of bantustan land and continue to charge rent for land access. Over time they have also been involved in a process of land alienation and resettlement where petit-bourgeois members of the community are further empowered while others are socially and spatially marginalized (see Figure 9.2). Participatory GIS can thus help to display the resource contradictions associated with the generally unpopular institution of the chieftaincy. Social differentiation, however, does not preclude the existence of community consensus on some issues, as the discussion on perceptions of land potential illustrates.

Second, another aspect of the mental map is the community perception of ecological space. A map composite of mental images, land types information obtained from the Department of Agriculture, and hydrology points to the value of linking local knowledge with traditional GIS data (see Figures 9.1 and 9.2). Most of the participants concur that KaNgwane soils are generally inferior to those found on adjacent white farms, and in the extreme western portion of the Kruger Park, although areas of higher quality soils were identified within the bantustan. For example, there was a general consensus in the workshop meetings that land along the four major perennial rivers is where the highest quality soils are located. Persons forcibly removed from land near rivers were particularly adamant regarding this point. The two knowledge sources both show surface water and precipitation availability as being important. The process of discussing and constructing the mental map uncovers additional information that is central for the successful implementation of rural land reform. For example, blacks once inhabited the major river valleys in the area and produced a diversity of crops. The process of forced removals destroyed indigenous livelihood systems and made them dependent on wage labor and imported food. During the apartheid era, the bantustan borders were in places moved away from the major perennial rivers, further reducing community access to water resources. Most blacks still cannot use river water and may get shot if they do; it was alleged that two boys were killed in 1991 because they were seen fishing in the Sand River. The political ecology of Kiepersol remains a core element of community consciousness and activism and materially drives the hardships of everyday life.

Although the inclusion of local knowledge and social information are at the core of our participatory methodology, conventional environmental and land use data are clearly also of great importance. Local knowledge does force decision makers, however, to question the validity and criteria of "expert" knowledge obtained as well as the basis for community evaluation of land potential and constraint. Evaluating land potential is very complex. Some inconsistencies in the representations of ecological space (see Figure 9.2) could be a reflection of scale. There is also the question of evaluating land potential in the context of specific land uses and the likelihood that community perspectives on land type and land potential are influenced by limited access to technology and peoples' everyday experiences with land overuse. Figures 9.1 and 9.2 thus demonstrate the utility of integrating differing forms of knowledge into a GIS, rather than accepting the superiority or inferiority of one form over another. It also points to the difficulties in the digital representation of local knowledge.

Third, the use of cognitive data in a GIS is associated with a number of technical issues that are not easily addressed. Current GIS technology is spatially deterministic. To an extent, an orientation toward the recording of physical environmental data within a GIS does hold out a concern for the return of GIS-derived environmental determination. Our attempts to incorporate sociocultural–perceptual data within a GIS seeks to redress this imbalance. However, cognitive information is geographically imprecise and is not expressed comfortably within a point/line/polygon paradigm. This emphasis on fuzziness also runs counter to the GIS focus on quantification and error minimization or management, and points to the importance of recent work on incorporating fuzziness within GIS.

Finally, because of the nature of current GIS technology, community knowledge transfer into a GIS must be filtered by outsiders, such as ourselves. This "knowledge distortion" necessarily separates the local community from their own cognitive information. The nuances of the local context, which underpins local knowledge, may easily be lost in the encoding process.

CONCLUSION

A number of issues have been raised in this chapter concerning the role of GIS in society and human geography and its potential utility in supporting democratic social transformation in South Africa. The geographic basis of apartheid emphasizes the importance of differential access to available spatial information and the political imbalances which access to GIS technology can reinforce. The scramble for data currently taking place in South Africa indicates the selective nature of data availability and the importance in which GIS is held by all parties as representing an advantaged position in preparing for the recent elections and developing postapartheid policy initiatives. In transitional South Africa, it is likely that GIS will reinforce traditional market-based and technicist approaches to policy formulation to the detriment of alternative restructuring strategies. This process is inherently undemocratic and is what Taylor, Pickles, and others have warned against. Our preliminary work addresses some of these concerns and attempts to counter the imposition and legitimation of traditional developmentalism for rural land reform in South Africa. This is done through the incorporation of local knowledge into a GIS through a participatory research and policy formulation process made possible by working with popular political structures within KaNgwane villages. This effort is at the core of the debate concerning the (un)democratic nature of GIS and the "ethnocentric universalism" assigned to traditional developmentalism.

Participation in itself cannot guarantee a democratic redressing of South Africa's historical political ecology. This is because rural communities in South Africa, as elsewhere, are not homogeneous entities. The question of "who" participates will be central to the outcome of a participatory process. Field experience has demonstrated clearly that it is the powerful (i.e., men, rich peasants, chiefs, and headmen) who tend to dominate participatory processes. While all of these social categories are part of the oppressed majority, oppression in South Africa is experienced differentially. Democratization must, therefore, be a transformative process that entails the development of new social relations which are likely to threaten certain local interests. In order to incorporate a GIS successfully, participation will have to be broad-based, inclusive, gender-sensitive, and biased toward the interests of marginalized people. Stratification of groups in workshops cannot ensure these sorts of biases will be overcome. If participatory GIS is genuinely to empower the most oppressed social segments of a community, then it must become part of a political program explicitly aimed at restructuring social relations, and hence contesting local and state political structures.

The use of GIS to support local empowerment through community participation is thus fraught with dangers and potential contradictions. While attempts to implement community input into participatory GIS are laudable, such efforts in our work are still dependent on advanced technology and expertise as well as databases created by the former institutions of apartheid. GIS also tends to rationalize knowledge from data and is biased toward technical interpretations of meaning. Thus in pursuing participatory GIS in South Africa, it is difficult to move beyond the type of top-down elitist process that the project seeks to redress. For the democratic use of GIS, it is important to educate local individuals and groups in the use of GIS, and its underlying fundamental geographical principles, and to provide access to the technology and data. In a country where access to even basic education was deliberately withheld for the majority of people, this represents a considerable hurdle which exposes some of the potential contradictions associated with participatory GIS. However, the importance of involving poor rural communities in identifying appropriate research questions and policies, as well as representing local knowledges, in a GIS should not be underestimated. In our experience, the mere process of involvement was empowering to the victims of apartheid with whom we are working. In the context of South Africa's current urban-biased political economy and policy environment, this in itself is an important and difficult achievement that is already yielding political results nationally.

The technical process of transforming local knowledge into spatial data raises additional questions. The term "local knowledge" requires very precise definition for it is problematic to assume that there is a homogeneous knowledge base. Socially differentiated knowledge may pose even greater problems for this effort than resource issues raised by differential access. Furthermore, one of the main planks underlying GIS is its ability to facilitate and support decision making. This has primarily been based on the notion of one single objective knowledge. With the inclusion of a combination of more than "one knowledge," it is likely that a GIS database will contain conflicting information and substantive fuzziness. Such was found to be the case in our study with the many differing interpretations by "experts," local participants, and the study team, as to the agroecology of Kiepersol. With the inclusion of locationally fuzzy local knowledge many issues begin to arise as to how multiobjective goals, based on multiple criteria, and using spatially imprecise and possible conflicting data, might actually achieve what is assumed to be consensus decision making. Indeed, greater quantities of information may promote social conflict. Perhaps one reason why GIS has achieved such astounding "success" to date in decision-making support roles is that it is based on only one seemingly noncontradictory perception of reality. It is in this spirit of combining physical data, socio-political information, and indigenous knowledge for the purposes of helping to redress the tragedy of apartheid that a participant GIS research methodology is being developed. As with much research, the project is raising as many issues regarding the role of GIS in society as it is answering.

ACKNOWLEDGMENTS

Financial support from the following sources is gratefully acknowledged: a West Virginia University Senate Grant, the West Virginia University Office of International Programs, and The John D. and Catherine T. MacArthur Foundation ("Community Perspectives on Land and Agrarian Reform in South Africa," grant #91-16564a, comanaged by R. Levin and D. Weiner). Thanks to Professor Edward Makhanya of the University of Zululand for help in obtaining the SPOT data, to Dr. Lilo de Gasparis and Anton Bloem of Anglo American Corporation for printing the satellite image, and to Dr. Phil Woodhouse for assistance obtaining and analyzing the land type information.

NOTES

1. This was the provocative title of a special session held at the Annual Association of American Geographers meeting in San Diego, April 18–21, 1993.

Organized and chaired by Lakshman Yapa and Ben Wisner, the session generated a very important discussion on the meaning of development and the role of technology.

2. The terms "African," "Colored," "White," and "Indian" are used when specific statistics or contexts warrant such usage, otherwise the term "black" is employed to indicate the population that is not white. This is the preferred term within the South African liberation movement.

REFERENCES

African National Congress. (1992). *Ready to govern: ANC policy guidelines for a democratic South Africa.* Johannesburg: Author.

Benmouffok, D. (1993). Information for decision making. *IDRC Reports, 20*(4), 4–5.

Brand, S., Christodoulou, N., van Rooyen, J., & Vink, N. (1992). Agriculture and redistribution: Growth with equity. In R. Schrire (Ed.), *Wealth or poverty? Critical choices for South Africa* (pp. 353–375). Cape Town: Oxford University Press.

Chief Director of Surveys and Mapping. (1984). *2531AA Kiepersol: 1:50,000 topography map.* Mowbray, South Africa: Trigonometric Survey.

Chrisman, N. R. (1987). Design of geographic information systems based on social and cultural goals. *Photogrammetric Engineering & Remote Sensing, 53*(10), 1367–1370.

de Janvry, A., Sadoulet, E., & Thorbecke, E. (1993). Introduction to Special Issue. *World Development, 21*(4), 565–575.

Edney, M. H. (1991). Stategies for maintaining the democratic nature of geographic information systems. *Papers and Proceedings of the Applied Geography Conferences, 14,* 100–108.

Escobar, A. (1992, June). Grassroots approaches and alternative politics in the third world. *Futures,* pp. 411–436.

Esteva, G. (1992). Development. In W. Sachs (Ed.), *The development dictionary: A guide to knowledge as power* (pp. 411–436). London: Zed Press.

Goodchild, M. F. (1991). Just the facts. *Political Geography Quarterly, 10*(4), 335–337.

International Bank for Reconstruction Development. (1993). *South Africa: Preparation of a rural restructuring program.* Unpublished manuscript. Washington, DC: World Bank.

Illich, I. (1985). The information age. *Mother Jones.*

Land Type Survey Staff. (1989). *Land types of the map 2530 Barberton. Memoirs on the agricultural natural resources of South Africa No. 13.* Pretoria, South Africa: Author.

Levin, R., & Weiner, D. (1993). The Agrarian question and politics in the "new" South Africa. *Review of African Political Economy, 57,* 29–45.

Long, N. (1992). Introduction. In N. Long & A. Long (Eds.), *Battlefields of knowledge: The interlocking of theory and practice in social research and development* (pp. 3–15). New York: Routledge.

Mabin, A. (1991). The impact of apartheid on rural reas of South Africa. *Antipode, 23*(1), 33–46.

MacKenzie, F. (1992). Development from within? The struggle to survive. In D. R. Fraser-Taylor & F. Mackenzie (Eds.), *Development from within* (pp. 1–32). London: Routledge.

Openshaw, S. (1991). A view on the GIS crisis in geography, or, Using GIS to put Humpty-Dumpty back together again. *Environment and Planning A, 23,* 621–628.

Pickles, J. (1991). Geography, GIS, and the surveillant society. *Papers and Proceedings of Applied Geography Conferences, 14,* 80–91.

Pickles, J., & Weiner, D. (1991). Rural and regional restructuring of apartheid: Ideology, development policy and the competition for space. *Antipode, 23*(1), 2–32.

Pickles, J., & Woods, J. (1992). South Africa's homelands in the age of reform: The case of QwaQwa. *Annals of the Association of American Geographers, 82*(4), 629–652.

Platzky, L., & Walker, C. (1985). *The surplus people: Forced removals in South Africa.* Johannesburg: Ravan Press.

Pred, A. (1992). Capitalisms, crises, and cultures, part 2: Notes on local transformation and everyday cultural struggles. In A. Pred & M. Watts (Eds.), *Reworking modernity* (pp. 106–117). New Brunswick, NJ: Rutgers University Press.

Pred, A., & Watts, M. J. (1992). *Reworking modernity: Capitalisms and symbolic discontent.* New Brunswick, NJ: Rutgers University Press.

Sachs, W. (Ed.). (1992). *The development dictionary: A guide to knowledge as power.* London: Zed Press.

Slater, D. (1992). On the borders of social theory: Learning from other regions. *Environment and Planning D: Society and Space, 10,* 307–327.

Smith, D. M. (1992). (Ed.). *The apartheid city and beyond: Urbanization and social change in South Africa.* London: Routledge.

South African Institute of Race Relations. (1993). *Race Relations Survey.* Johannesburg: Author.

Taylor, P. J. (1990). GKS. *Political Geography Quarterly, 3,* 211–212.

Taylor, P. J. (1991). A distorted world of knowledge. *Journal of Geography in Higher Education, 15,* 85–90.

Taylor, P. J., & Overton, M. (1991). Further thoughts on geography and GIS. *Environment and Planning A, 23,* 1087–1094.

Watts, M. J. (1993). Development, part 1: Power, knowledge, discursive practice. *Progress in Human Geography, 17*(2), 257–272.

Weiner, D. (1991). Socialist transition in the capitalist periphery: A case study of agriculture in Zimbabwe. *Political Geography Quarterly, 10*(1), 54–75.

Weiner, D., & Naidoo, D. (1993). Transforming Bantustan agriculture in South Africa. In M. D. Nellis (Ed.), *Geographic perspectives on the social and economic restructuring of rural areas: Proceedings of the IGC Commission on Changing Rural Systems* (pp. 14–26). Manhattan: University of Kansas Press.

Wilson, F., & Ramphele, M. (1989). *Uprooting poverty: The South African challenge.* Cape Town: David Phillip.

Woodhouse, P. (1994). *Soils and irrigation systems in the Hazyview area of the Eastern Transvaal.* (Working paper for the project entitled "Community Perspective on Land and Agrarian Reform in South Africa"). Johannesburg: University of the Witwatersrand.

Yapa, L. S. (1991). Is GIS appropriate technology? *International Journal of Geographical Information Systems, 5,* 41–58.

CHAPTER 10

Conclusion

TOWARD AN ECONOMY OF ELECTRONIC REPRESENTATION AND THE VIRTUAL SIGN

John Pickles

[Even] . . . if Virtual Reality technology never lives up to its promise, the phrase will continue to take root in the common language, signalling a widely-held confusion about how we encounter the objects, images and ideas around us. (Rosenthal, 1992, p. 107)

One of the foremost tasks of art has always been the creation of a demand which could be fully satisfied only later. The history of every art form shows critical epochs in which a certain art form aspires to effects which could be fully obtained only with a changed technical standard, that is to say, in a new art form. (Benjamin, 1968, p. 237)

As telecommunications, information processing, and spatial imaging systems have emerged as everyday realities, and as geographic information systems (GIS) have been incorporated into planning, design, and marketing practices, questions about their impacts have become more pressing. In particular, questions are increasingly being asked about the potential uses GIS opens up for us as geographers, the impacts of GIS on disciplinary concepts and practices, and the ability of enhanced imaging capabilities to improve how we make decisions. But it is also clear that electronic data handling and imaging software and hardware are big business, and that developers and professional users have become embedded in mutual relations of benefit and need. Consumers for these new products have been painstakingly produced, and they now operate as collaborators with producers to sustain and expand demand. These issues raise further questions: How will system-lifeworld relations be mediated under regimes of electronic accounting, management, and mapping? How will the colonization of everyday life by systems of "management" be enhanced or hindered by information-mapping technologies? How will the new telematics and informatics contribute to the projects of state monitoring, capital penetration, and military enclosure? How will class politics and a politics of difference fare under electronic administration and new representational practices? And, what new capacities are emerging to sustain local, antihegemonic discourses and practices?

Yet, despite the rapid adoption and widespread use of new forms of "technocartography" and "surveillant technologies" of knowledge engineering, no sustained critique of the role of surveillance within science and in the practices of state and private agencies has yet emerged within geography (but see Pickles, 1991; Smith, 1992; O'Tuathail, 1993). This is surprising, given the prior, and in many ways parallel, critique of technicist approaches to science that arose in the 1970s and the rapidity with which extended and new forms of surveillance have been adopted since then.

The authors in this collection have engaged GIS at several levels. Some have focused on the possibilities for enhanced data handling and map production offered by recent developments in GIS. Others have focused their attention on the ways in which disciplinary concepts and practices are being reconfigured under its influence. For still others, the task has been to write a political economy of technology, a social theory of the virtual/electronic sign, and/or a disciplinary history of GIS and postmodern excess. Several of the authors have turned variously to Lefebvre, Foucault, de Certeau, Virilio, and others to help them describe and theorize the impacts of new electronic ways of world-making and contribute to what I referred to in Chapter 1 as new ways of "wordmaking."

In this concluding chapter I will map out three possible visual imaginaries, each of which raises questions about future paths for charting a critical geography and social theory of GIS. I make no attempt here to integrate or elide the disjunctures between each scripting, but pose them as possibilities that emerge from the issues raised by the authors in the book and that ask us to push further our autocritique of our new graphisms. They are the emergence of new representational economies based on systems of digitality; locality and community politics in an age of delocalization and generalized "community" identity; and the democratic appropriation of digital, spatial technologies.

DIGITALITY, REPRESENTATION, AND INTERTEXTUALITY

> *Just as water, gas, and electricity are brought into our houses from far off to satisfy our needs in response to a minimal effort, so we shall be supplied with visual or auditory images, which will appear and disappear at a simple movement of the hand, hardly more than a sign. (Valery, 1964, p. 226)*

Valery's understanding of the coming of a new age of information and images, or what Gianni Vattimo (1992, p. 1) has called a society of generalized communication, points to the importance of the current debate about GIS, geography, and democratic practice. In this postmodern world, "the world of generalized communication explodes like a multiplicity of 'local' rationalities—ethnic, sexual, religious, cultural or aesthetic minorities—that finally speak up for themselves. They are no longer repressed and cowed into silence by the idea of a single true form of humanity that must be realized irrespective of particularity and individual finitude, transience and contingency" (Vattimo, 1992, p. 9). The revolution in telematics, informatics, and virtual representation has—in this view—resulted in the dissolution of "grand narratives" (Lyotard) and "centralized perspectives" (Vattimo). One consequence of this world of generalized communication and partial discourses is that

> the self-transparency to which we are at present being led by the ensemble of media and human sciences seems to be nothing more than the exposure of pluralism, of the mechanisms and inner fabric of our culture. Even at its best, the media-human sciences complex is emancipatory only inasmuch as it places us in a world less univocal, less certain and so also much less reassuring than that of myth. (Vattimo, 1992, p. 26)

For Walter Benjamin (1968, p. 221), the coming of an age of mechanical reproduction resulted in the withering of the aura of the work of art: "The technique of reproduction detaches the reproduced object from the domain of tradition. By making many reproductions it substitutes a plurality of copies for a unique existence. And in permitting the reproduction to meet the beholder or listener in his own particular situation, it reactivates the object reproduced." With the emergence of GIS and other spatial imaging systems, the age of mechanical reproduction (and particularly the age of the camera) is giving way to an age of mechanical hyperreproduction (Xerox machine, computer printer) conjoined with a world of infinite manipulability (computerized digital imagery). In this world of electronic representation the auratic nature of the fixed-object image is destabilized in favor of a playful and willful reproducibility and manipulability. Such forms of electronic reproduction have the character of resituating and reconfiguring "the object" (the earth, the region, land parcels) in multiple contexts and in ways that call for new theories of maps, landscape, and geography.

Digitality and representational technologies (such as GIS) produce new codings and practices, and with them new possible geographies. Whether in the secret labs of the U.S. Air Force, where new human-machine interfaces are being operationalized; in the studios of innovative television producers, where virtual realities and cyberspace are being "concretized" and rapidly disseminated as "real futures"; in ads for sports shoes, soft drinks, or political parties, where new cultural codings are occurring; in the operating theaters of hospitals, where digital and spectral imaging techniques are combined to produce real-time internal pictures of organs at work; or in planning offices or the lecture rooms of art departments, where students are being challenged to extend the visual powers of cyberart and virtual worlds, technologies of visual representation are challenging our fundamental categories of objectness, clear sight, the seen and the unseen, the obvious, the stable, and the exterior and interior.

The dislocation of universals and of the authority of the text is at the same time an opening up of the intertextuality of all texts. GIS is a system of spatial data handling and representation, and as such incorporates multiply embedded systems of texts (including signs, databases, and representations).[1] All texts are, in this sense, embedded within chains of signification: meaning is dialogic, polyphonic, and multivocal —open to, and demanding of us, a process of ceaseless contextualization and recontextualization. Intertextuality cannot be fused with positivist or more broadly empiricist epistemologies, but requires a thoroughly different understanding of epistemology—a rejection of the univocity of texts (and images), of representation as a mirror of nature, and of a metaphysics of presence (and the foundational claims of positivism) to ground itself unproblematically in the given real world or the immediacy of observation. These implications are devastating for so much of what stands for theory of mapping in contemporary GIS, which is so heavily dominated by what Jameson (1971, p. x) has called the "anti-speculative bias" of the liberal tradition with its "emphasis on the individual fact or item at the expense of the network of relationships in which that item may be embedded"—a bias which keeps people from "drawing otherwise unavoidable conclusions at the political level."

Intertextuality implies a decentering of the author and the reader, and the situating of meaning in the margins between texts and writers—in an illimitable chain of signification, a network that as Heinz Pagels (1989) suggests has "no 'top' or 'bottom' . . . no central executive authority that oversees the system." In this sense, the turn from the text as isolated object to a text embedded in a constantly expanding chain of signification could mean for GIS, in the way it means for Gerard Genette (1982, p. 147, quoted in Landow, 1992, p. 60), an escape from a "sort of idolatry, which is no less serious, and today more dangerous" than the idealization of the author, "namely, the fetishism of the work—conceived of as a closed, complete, absolute object." Moreover, implied in this shift to an understanding of texts as constantly open to interpretation and critique, is a different understanding of what texts are for. Roland Barthes (1975) has called this a change from "readerly texts" (whose purpose is to create readers for already written texts) to "writerly texts" (whose purpose is to see texts as producing an open series of readings, each of which requires that the reader also be in part the author of meaning).

The shift that occurs from the "tactile" (pen and ink) to the "digital" (electronic code) with the adoption of GIS combines fixity and flexibility in new ways (Landow, 1992, p. 19; see also Baudrillard, 1983, 1988; Kellner, 1989). In this sense, "digitality is with us. It is that which haunts all the messages, all the signs of our societies. But this digitality also brings with it a commitment to a binary logic connected to a particular metaphysical principle: cybernetic control . . . the new *operational* configuration" (Baudrillard, 1983, p. 103). The theorist of GIS sits with two apparently incommensurable "ontologies"—Derrida and Baudrillard, intertextuality and cybernetic control through simulation—paradoxical strands running through the question of virtuality and digitality, and requiring of GIS and geography a much fuller engagement with the theory of signs, dissimulation, and inscription.

DETERRITORIALIZATION, DELOCALIZATION, AND "COMMUNITY" IDENTITY

> *Communities are to be distinguished, not by their falsity/genuineness, but by the style in which they are imagined. (Anderson, 1983, p. 6)*

> *The "new utopia" presupposes the disappearance of the local in favor of the spatial. Delocalization means the insertion of the logic of the new communication technologies within the universal history understood as rationalization. . . . The technical utopia of a society decentralized by telecommunications signifies a spatialization of communication so that all localization becomes impossible. It means the final dissolution of all ties and places that symbolically structured traditional society. (Raulet, 1991, p. 40)*

Vattimo's intriguing flirtation with the emancipatory possibilities of the media-human sciences that began this discussion of digitality, and my own consideration of the intertextuality of GIS images, are, however, somewhat misleading. Decentering of the subject, the pluralization of culture, and GIS object as critically embedded text must also be seen against a backcloth of increasing monopoly control over GIS and other electronic technologies. As many of the chapters in this book have shown, GIS has become an important commodity and (along with hypertext, multimedia, and virtual reality) a very profitable frontier for investment. In this context of market ideology and monopoly capital, everything comes under the sway of "information" as an object of counting, measuring, and analysis. Modern technological society even sets up human beings and nature as objects of manipulation in such ways that "our whole human existence everywhere sees itself challenged—now playfully and now urgently, now breathlessly and now ponderously—to devote itself to the planning and calculating of everything" (Heidegger, 1957/1969, pp. 34–35). As a consequence, "what could freedom of information, or even the existence of more than one radio or TV channel, mean in a world where the norm is the exact reproduction of reality, perfect objectivity, the complete identity of map and territory?" (Vattimo, 1992, pp. 6–7).

Kroker (1992, pp. 1–2) has gone so far as to suggest that "refusing the pragmatic account of technology as freedom and eschewing a tragic description of technology as degeneration, an arc of twentieth century French thinkers, from Jean Baudrillard and Roland Barthes, to Paul Virilio, Jean-Francois Lyotard, Deleuze and Guattari and Foucault have presented a description of technology as cynical power. . . . Here, technological society is described under the sign of possessed individualism: an invasive power where life is enfolded within the dynamic technological language of virtual reality.''

Certainly, the ways in which new electronic technics create new conditions for domination and the exercise of power were at the heart of Lyotard's (1984) report on knowledge in *The Postmodern Condition:*

> Our working hypothesis is that the status of knowledge is altered as societies enter what is known as the postindustrial age and cultures enter what is known as the postmodern age. This transition has been under way since at least the end of the 1950s. . . .
>
> I will take as my point of departure a single feature, one that immediately defines our object of study. Scientific knowledge is a kind of discourse. And it is fair to say that for the last forty years the "leading'' sciences and technologies have had to do with language: phonology and theories of linguistics, problems of communication and cybernetics, modern theories of algebra and informatics, computers and their languages, problems of translation and the search for areas of compatibility among computer languages, problems of information storage and data banks, telematics and the perfection of intelligent terminals. (p. 3–4)

Lyotard (1984) saw the effects of these changes as deep and serious:

> It is reasonable to suppose that the proliferation of information-processing machines is having, and will continue to have, as much of an effect on the circulation of learning as did advancements in human circulation (transportation systems) and later, in the circulation of sounds and visual images (the media). (p. 4)

At the end of his report on the condition of knowledge Lyotard (1984) leaves us with this warning:

> We are finally in a position to understand how the computerization of society affects this problematic. It could become the "dream'' instrument for controlling and regulating the market system, extended to include knowledge itself and governed exclusively by the performativity principle. In that case, it would inevitably involve the use of terror. But it could also aid groups discussing metaprescriptives by supplying them with the information they usually lack for making knowledgeable decisions. The line to follow for computerization to take the second of these two paths is, in principle, quite simple: give the public free access to the memory and data banks. Language games would then be games of perfect information at any

> given moment. But they would also be non-zero-sum games, and by virtue of that fact discussion would never risk fixating in a position of minimax equilibrium because it had exhausted its stakes. For the stakes would be knowledge (or information, if you will), and the reserve of knowledge—language's reserve of possible utterances—is inexhaustible. This sketches the outline of a politics that would respect both the desire for justice and the desire for the unknown. (p. 67)

David Harvey (1989, p. 117) rejects as "silly" Lyotard's prescription for opening up the data banks to everyone as a condition for radical reform, and instead argues that this notion of postmodernity represents a dangerous shift from ethics and political commitment to aesthetics and a new "structure of feeling." In *The Condition of Postmodernity* Harvey shows how informatics operate as forms of power and how information systems are at the heart of a celebratory postmodernism which has emerged out of antimodernism as a "cultural aesthetic in its own right" (Harvey, 1989, pp. 3, 49). The aestheticization of information and information systems operates through the fetishization of technology and representations themselves. Thus, Lyotard's postmodern interpretation fails to understand the ways in which changes in fashion (including technology, ideas, and representations) are reflective of changes in the economic and social fabric. In this sense postmodernism "signals nothing more than a logical extension of the power of the market over the whole range of cultural production" (Harvey, 1989, p. 62); the postmodern fascination with information is one element of the cultural logic of late capitalism (Jameson, 1984).

In the place of this politically dangerous and ethically nihilistic postmodern fascination, Harvey (1989) indicates that we must consider the ways in which electronic information systems emerged as a result of changes in the structure of capitalism and the liberal state, as each struggled to deal with the fiscal and legitimacy crises of the 1970s. In other words, we need a political economy of information and technology, and we need to see how each operates within the broader restructuring of late captalism. This restructuring is not technologically determined nor driven by markets, but it is part of a broader class struggle to create new relations of domination in the workplace, to put in place in production more efficient technical and organizational practices, to extend patterns of commodification and commercialization into new forms and new niches (including information and data), and to orchestrate new modes of social control and new methods of conducting war (see Clarke, 1988).

Thus, we need to ask again about the ways in which electronic information and mapping technologies are reconfiguring the contemporary world. As counting machines and typewriters had done earlier, new computerized information systems and artificial neural networks facilitate data entry, capture, and reproduction (Benjamin, 1968).[2] Informatics effect new capacities in speed, efficiency, and the reduction of effort by which we communicate and act (see Virilio & Lotringer, 1983; Virilio, 1986). These new forms of experience correspond in part to the shift from a modernist Fordism to a liberal productivism, postmodernism, and post-Fordism (Lipietz, 1992). They also emerge at the boundary of the Cold War, and here GIS (and related information-handling and imaging systems) functions to create new codes whose liminal futures and new geographies are yet to be written. Mapping techniques extend a rationalistic logic—a universal calculus—to unify space as object, material, and fundament, and earth as exploitable resource, unified community, or commercial logo. A naturalized present is scripted and inscribed within the domains of cultural production, in terms of which new cultural imaginaries of natural (earth, nature, globe) and social identity are being forged, and electronic images of the earth, interactions in cyberspace, conversations on the community net, or concrete engagements with virtual reality represent self and others in new ways, create alternative forms of experience, and establish new forms of social interaction. Fully normalized, the technics of data exchange and representation legitimize new social practices and institutions in ways that we have only begun to recognize and regulate.

Such reconfigurations directly affect the ways in which citizens and democratic practices operate at all scales, and raise important questions about whether we are building an electronic global village, global corporate monopolies, or new civic arenas for citizen action. In the process, democratic polities experience the universalization and freezing in place of particular institutional forms (e.g., representative, liberal democracy), and the power to visualize alternatives gives way to the power to assert the value of a single representation of freedom (Fukuyama, 1992; Peet, 1993a, 1993b). If Benedict Anderson (1983) is correct to point to the role of print capitalism in the emergence of imagined communities of hegemonic national identity, in what ways is electronic/computer capitalism (or to use Ben Agger's [1989] phrase, "fast capitalism") now building an image of militaristic internationalism and corporatist global government? "Living without alternatives"—as Zygmunt Bauman (1992) expresses the contemporary situation—is also living with systems of information gathering, data manipulation, and image production that naturalize the present ideology of commodification, restructuring, and creative destruction.

At home and abroad, governments like the Bush administration script and inscribe new versions of state power, in which information imaging systems facilitate military, political, and economic goals simultaneously. Thus, the Persian Gulf War was the first GIS war, although only the latest in a long line of geographical wars (Clarke, 1992). It must be remembered that General Powell was one of the Pentagon's top experts on electronic information systems and smart systems, and Robert Hanke (1992, p. 136) has pointed out how in the Persian Gulf War smart weaponry, GIS technology, and telecommunications (including CNN) were carefully orchestrated and coordinated—a "kind of simulacra game in which the technology of entertainment television and the technology of mass destruction were deployed together as part of U.S. military strategy to both deceive Iraqi military forces and to pre-empt/post-empt the formation of an oppositional public sphere."

Even this combination of militaristic goals with the "need" to manage public opinion at home must be further contextualized in terms of the economic conditions to which both are related:

> It is thanks to the financial leverage provided by this firm commitment to the warfare state that the corporate community has been able to engineer the wrenching break with America's industrial past we now find ourselves in. In large part, the advent of the information economy means that our major corporations are rapidly retiring two generations of old capital or moving it abroad. As they do so, with the rich support of military contracts, they are liberating themselves from the nation's most highly unionized labor so that investment may be transferred into more profitable fields. High tech is not only glamorous; it pays off handsomely, especially if those who are collecting the profits are excused from paying the social costs that result from running down old industrial centers and disempowering their work force. (Roszak, 1986, p. 28)

New methods of data handling and electronic imaging are producing deep shifts in the ontology of modern life, and specifically changes in the role of the visual in the contemporary world: As Jameson (1992,

p. 1) has argued, ''All the fights about power and desire have to take place here, between the mastery of the gaze and the illimitable richness of the visual object.'' And it is precisely here—in the intersection of the mastery of the gaze and the textual malleability of electronic images—that geographic theory remains surprisingly silent, particularly about the ways in which GIS has begun to effect deep-seated changes in the discursive practices of the discipline, the broader economy of information, and the uses to which such imaging techniques are put.

At the same time, as the dichotomy between spheres of production and consumption, and between productive and unproductive social practices, has collapsed, there has emerged a tension between universalized administrative rationalities and the expansion of technologies in which subjects are constituted as objects of economic and political manipulation *and* a process of subjectivation in which the potential exists for an ''expansion of the categories of who or what can be a political subject'' (Feldman, 1993, p. 6). Thus, the totalization of information and the hegemonic projects of monopoly and bureaucracy take place within a system of cultural production characterized by an ''apparently irresistible pluralization [which] renders any unilinear view of the world and history impossible'' (Vattimo, 1992, p. 6). From this perspective, the very inconstancy and superficiality of experience reflected in the media-human sciences opens the possibility that generalized communication, mass media, and reproduction do not lead only to standardization, uniformity, and the manipulation of consensus. The contingent nature of political and social outcomes which follow from the incorporation of technologies from and in the media-human sciences challenge any straightforward functionalist and reductive interpretation of technology as constraint, and challenge us to think further about the democratizing possibilities of GIS, as well as the ways in which GIS must be changed if it is to function as other than an instrument of power and control.

DEMOCRATIZING APPROPRIATIONS?

What is the social basis for the contemporary decay of the aura of the image-object?

It rests on two circumstances, both of which are related to the increasing significance of the masses in contemporary life. Namely, the desire of contemporary masses to bring things

> *"closer" spatially and humanly, which is just as ardent as their bent toward overcoming the uniqueness of every reality by accepting its reproduction. Every day the urge grows stronger to get hold of an object at very close range by way of its likeness, its reproduction. Unmistakably, reproduction as offered by picture magazines and newsreels differs from the image seen by the unarmed eye. Uniqueness and permanence are as closely linked in the latter as are transitoriness and reproducibility in the former. To pry an object from its shell, to destroy its aura, is the mark of a perception whose "sense of the universal equality of things" has increased to such a degree that it extracts it even from a unique object by means of reproduction. Thus is manifested in the field of perception what in the theoretical sphere is noticeable in the increasing importance of statistics. The adjustment of reality to the masses and the masses to reality is a process of unlimited scope, as much for thinking as for perception. (Benjamin, 1968, p. 223)*

The unlimited scope for the "democratization" of the image and information for the masses, and the corresponding adjustment of the masses to this new reality, have important implications for our understanding of the democratic potential of GIS and electronic technology generally. Up to now, the debate about the implications of new electronic technologies has tended to ossify around a comfortable dualism across whose borders insults and occasional mortars are lobbed, but between whose positions little that passes for sustained critique has emerged.[3] On the one side, GIS is claimed to enhance access to information and therefore potentially (and inherently) can be used to enhance democratic practices. On the other side, GIS is seen to foster the interests of particular users and to produce increasingly constrained and controlled public spheres.

Of course, since the social changes taking place are not restricted to the impacts of GIS/electronic/computer technologies, but reflect a broader restructuring of social relations, the debate cannot be limited only to GIS and geography. The issue is one with a much longer history, as Benjamin (1968) has shown us:

> With the increasing extension of the press, which kept placing new political, religious, scientific, professional, and local organs before the readers, an increasing number of readers became writers—at first, occasional ones. It began with the daily press opening to its readers space for "letters to the editor." And today there is hardly a gainfully employed European who could not, in principle, find an opportunity to publish somewhere or other comments on his work, grievances, documentary reports, or that sort of thing. Thus, the dis-

> tinction between author and public is about to lose its basic character. The difference becomes merely functional; it may vary from case to case. At any moment the reader is ready to turn into a writer. (p. 232)

What are the effects of these changes? On the one hand, the technologies of reproducibility can—in the hands of a Brecht—be turned into critical theater that challenges and destabilizes the categories and arrogance of bourgeois culture and life. Or like the letter writer, the pc user at home has the opportunity to use the edifice of complex information processing and mapping technologies as he or she wishes. The individual engages the system, and molds it to his or her particular needs. On the other hand, technoculture redefines even the notion of who this individual agent is: as, for example, when GIS applications constitute new individual identities based on neighborhood data profiles and aggregate characteristics. These "produced identities" are deemed sufficiently "accurate" insofar as they facilitate "effective" action, a form of instrumentalism that is proliferating in many spheres of GIS use, whether the object is a parcel of land, terrain for military action, or a neighborhood of "like" consumers.

GIS technologies and practices are increasingly being used (along with other abstractive and distancing technologies) to produce collective identities. The consequences can be serious, as Benjamin (1968) has shown us:

> Mass reproduction is aided especially by the reproduction of masses. In big parades and monster rallies, in sports events, and in war, all of which nowadays are captured by camera and sound recording, the masses are brought face to face with themselves. This process, whose significance need not be stressed, is intimately connected with the development of the techniques of reproduction and photography. Mass movements are usually discerned more clearly by a camera than by the naked eye. A bird's-eye view best captures gatherings of hundreds of thousands. And even though such a view may be as accessible to the human eye as it is to the camera, the image received by the eye cannot be enlarged the way a negative is enlarged. This means that mass movements, including war, constitute a form of human behavior which particularly favors mechanical equipment. (p. 251)

According to Benjamin (1968, p. 241), the "proletarianization of modern man and the increasing formation of masses are two aspects of the same process. Fascism attempts to organize the newly created proletarian masses without affecting the property structure which the masses strive to eliminate. Fascism sees its salvation in giving these masses not their right, but instead a chance to express themselves; Fascism seeks to give them an expression while preserving property." The result is the aestheticization of political life and the celebration of ritual values.

> All efforts to render politics aesthetic culminate in one thing: war. War and war only can set a goal for mass movements on the largest scale while respecting the traditional property system. This is the political formula for the situation. The technological formula may be stated as follows: Only war makes it possible to mobilize all of today's technical resources while maintaining the property system. (Benjamin, 1968, p. 241)

Benjamin's discussion of the aestheticization of war under fascism and Marinetti's manifesto on the Ethiopian colonial war have an eerie parallel in the emergence of new technologies of war in the late 20th century. In particular, the incorporation of GIS technologies in the Persian Gulf War "pinpoint" blanket bombing of Iraq, its presentation through television, and its aestheticization in a ritual celebration of national power and virtue, raise important questions about the nature of new information technologies and their claims to foster democratic practices.

> For twenty-seven years we Futurists have rebelled against the branding of war as antiaesthetic. . . . Accordingly we state . . . War is beautiful because it establishes man's domination over the subjugated machinery by means of gas masks, terrifying megaphones, flame throwers, and small tanks. War is beautiful because it initiates the dreamt-of metalization of the human body. War is beautiful because it enriches a flowering meadow with the fiery orchids of machine guns. War is beautiful because it combines the gunfire, the cannonades, the cease-fire, the scents, and the stench of putrefaction into a symphony. War is beautiful because it creates new architecture, like that of the bog tanks, the geometrical formation flights, the smoke spirals from burning villages, and many others. . . . Poets and artists of Futurism! . . . remember these principles of the aesthetics of war so that your struggle for a new literature and a new graphic art . . . may be illumined by them! (Marinetti, quoted in Benjamin, 1968, pp. 241–242)

Benjamin's (1961) response to this way of thinking is straightforward and clear:

> If the natural utilization of productive forces is impeded by the property system, the increase in technical devices, in speed, and in the sources of energy will press for an unnatural utilization, and this is found in war. The destructiveness of war furnishes proof that society has not been mature enough to incorporate technology as its organ, that technology has not been sufficiently developed to cope with the elemental forces of society. The horrible features of imperialistic warfare are attributable to the discrepancy between the tremendous means of production and their inadequate utilization in the process of production—in other words, to unemployment and the lack of markets. Imperialistic war is a rebellion of technology which collects, in the form of "human material," the claims to which society has denied its natural material. Instead of draining rivers, society directs a human stream into a bed of trenches; instead of dropping seeds from airplanes, it drops incendiary bombs over cities; and through gas warfare the aura is abolished in a new way. (p. 242)

Questions about democratic possibilities and threats to democracy are thus relocated as a question about the nature of technology as a social relation, and in particular the question of what is its relationship to the property system?

We began this book with a disciplinary debate about a disciplinary object—GIS—and we have continued and extended that discussion to take into account the ways in which GIS as object, research enterprise, disciplinary practice, commodity, technology, imaging system, and social practice also operates as a disciplining object, full of new potentialities, both restrictive and constitutive. The artificiality of closure that typifies much debate about GIS (whether as technical or disciplinary limit) must be overcome if geographers are to begin to debate GIS itself as object—that is, as discursive and nondiscursive practices and institutions, which are centrally involved in the deployment of new modes of power and new capacities in the contemporary world:

> The challenge to understand a social world so profoundly mediated by simulation, to understand it—let's say to help bring about liberatory activity within it—is the daunting task that awaits anybody trying to examine the claims of the technologies of simulation, and the postmodern theories that attend them. (Rosenthal, 1992, p. 116)

But to deepen our understanding of the impact of GIS as technology, object, practice, and social relation, it will be necessary to broaden the context within which the disciplinary history is written. The chapters in this book begin this process, but the discussion must be carried much further if the pace of technological incorporation is in any way to preserve, or open up, the possibilities for democratic discourse and action. The alternative—as Benjamin saw so clearly—is an aestheticizing of technology's public practices and effects, and a "liberalizing" and instrumentalizing of the academic reception of the new spatial information and imaging technologies. In both contexts we continue to need a critical theory of social life, and a much stronger debate about democracy, property, individual rights and responsibilities, the limits of state power, and the exercise of competition. As the capacities and applications of spatial data and imaging systems continue to be broadened and deepened by forces of cybernetic capitalism and the celebration of technoscience, these questions remain pressing and open. We are only at the beginning of the process of delimiting and mapping the territory and content of this new currently solidifying terra incognita.

NOTES

1. These few paragraphs on digitality and intertextuality rely on themes developed in more detail in George Landow's (1992) *Hypertext: The Convergence of Contemporary Critical Theory and Techology.* For a discussion of the notion of text implied here, see Derrida (1986, p. 366): "*Text,* as I use the word, is not the book. No more than writing or trace, it is not limited to the *paper* which you cover with your graphism. Its precisely for strategic reasons . . . that I found it necessary to recast the concept of text by generalizing it almost without limit, in any case without present or perceptible limit, without any limit that *is.* That's why there is nothing '*beyond* the text.' "

2. Benjamin (1968) suggested that the new often enters on to the stage of history presented in the guise of its predecessor. Thus, the train emerges in the guise of a roaring bull, and the computer emerges as, and in the form of, a typewriter or a counting machine. The struggle over DOS-based and MAC-based platforms reflects this struggle over the representation of the new. The enormous success and apparent messianic devotion of MAC users to the pictorial form of access to operations in the MAC system contrast with the rather more stolid refusal of DOS-based users (of which I was one) to accept in the place of linguistic commands new representational forms. The transition into a new age of representations is marked by the rapid emergence and enormous success of Microsoft Windows for DOS users. Slowly the dinosaurs are incorporated (even aganst their will) into the age of electronic representations.

3. For a broader critique of the influence of paradigmatic approaches to science and debate which underpins this situation, see Pickles and Watts (1993).

REFERENCES

Agger, B. (1989). *Fast capitalism: A critical theory of significance.* Urbana: University of Illinois Press.

Anderson, B. (1983). *Imagined communities: Reflections on the origins and spread of nationalism.* London: Verso.

Barthes, R., (1975). *S/Z* (R. Miller, Trans.). New York: Hill and Wang.

Baudrillard, J. (1983). *Simulations* (P. Foss, P. Patton, & P. Beitchman, Trans.). New York: Semiotext(e).

Baudrillard, J. (1988). *Jean Baudrillard: Selected writings.* (M. Poster, Ed. & Trans.). Stanford, CA: Stanford University Press.

Bauman, Z. (1992). *Intimations of postmodernity.* New York: Routledge.

Benjamin, W. (1968). The work of art in the age of mechanical reproduction. In H. Arendt (Ed.), *Illuminations: Essays and reflections* (pp. 217–257). New York: Schocken Books. (Original work published 1931)

Clarke, K. C. (1992). Maps and mapping technologies of the Persian Gulf War. *Cartography and Geographic Information Systems, 19*(2), 80–87.

Clarke, S. (1988). Overaccumulation, class struggle and the regulation approach. *Capital and Class, 36,* 59–92.

Derrida, J. (1976). *Of grammatology.* (G. C. Spivak, Trans.). Baltimore: Johns Hopkins University Press.

Derrida, J. (1986). But, beyond . . . (P. Kamuf, Trans.). In H. L. Gates (Ed.). *"Race," writing, and difference* (pp. 354–369). Chicago: University of Chicago Press.

Feldman, A. (1993). *Formations of violence. The narrative of the body and political terror in Northern Ireland.* Chicago: University of Chicago Press.

Fukuyama, F. (1992). *The end of history and the last man.* New York: Avon Books.

Genette, G. (1982). Stendhal. In A. Sheridan (Ed.), *Figures of literary discourse.* New York: Columbia University Press.

Hanke, R. (1992). The first casualty? *Public, 6: Violence,* 134–140.

Harvey, D. (1989). *The condition of postmodernity: An inquiry into the origins of cultural change.* Oxford: Basil Blackwell.

Heidegger, M. (1969). *Identity and difference.* (J. Stambaugh, Trans.). New York: Harper and Row. (Original work published 1957)

Jameson, F. (1971). *Marxism and form: Twentieth century dialectical theories of literature.* Princeton, NJ: Princeton University Press.

Jameson, F. (1984). Postmodernism, or the cultural logic of late capitalism. *New Left Review, 146,* 53–92.

Jameson, F. (1992). *Signatures of the visible.* New York: Routledge.

Kellner, D. (1989). Media, simulations and the end of the social. In *Jean Baudrillard: From Marxism to postmodernism and beyond* (pp. 60–92). Stanford, CA: Stanford University Press.

Kroker, A. (1992). *The possessed individual: Technology and the french postmodern.* New York: St. Martin's Press.

Landow, G. P. (1992). *Hypertext: The convergence of contemporary critical theory and technology.* Baltimore: Johns Hopkins University Press.

Lipietz, A. (1992). *Towards a new economic policy.* Oxford: Oxford University Press.

Lyotard, J. F. (1984). *The postmodern condition: A report on knowledge.* (G. Bennington & B. Massumi, Trans.). Minneapolis: University of Minnesota Press.

O'Tuathail, G. (1993). The effacement of place? U.S. foreign policy and the spatiality of the Gulf Crisis. *Antipode, 25*(1), 4–31.

Pagels, H. R. (1989). *The dreams of reason: The computer and the rise of the sciences of complexity.* New York: Bantam Books.

Peet, R. (1993a). Reading Fukuyama: Politics at the end of history. *Political Geography, 12*(1), 64–78.

Peet, R. (1993b). The end of prehistory and the first man. *Political Geography, 12*(1), 91–95.

Pickles, J. (1991). Geography, GIS, and the surveillant society. *Papers and Proceedings of the Applied Geography Conferences, 14,* 80–91.

Pickles, J., & Watts, M. (1992). Paradigms for inquiry? In R. F. Abler, M. G. Marcus, & J. M. Olson (Eds.). *Geography's inner worlds: Pervasive themes in contemporary American geography* (pp. 301–326). New Brunswick, NJ: Rutgers University Press.

Raulet, G. (1991). The new utopia: Communication technologies. *Telos, 87,* 39–58.

Rosenthal, P. (1992). Remixing memory and desire: The meanings and mythologies of virtual reality. *Socialist Review, 22*(3), 107–117.

Roszak, T. (1986). *The cult of information: The folklore of computers and the true art of thinking.* New York: Pantheon Books.

Smith, N. (1992). Real wars, theory wars. *Progress in Human Geography, 16*(2), 257–271.

Strategic Simulations, Inc. (1992). Cyber-Empires [Advertisement]. *Computer Game Review and CD-ROM Entertainment, 2*(3), 83.

Valery, P. (1964). The conquest of ubiquity. In *Aesthetics.* (R. Manheim, Trans.). New York: Pantheon Books.

Vattimo, G. (1992). *The transparent society.* Baltimore: Johns Hopkins University Press.

Virilio, P. (1986). *Speed and politics: An essay on dromology* (M. Polizzotti, Trans.). New York: Semiotext(e).

Virilio, P., & Lotringer, S. (1983). *Pure war* (M. Polizzotti, Trans.). New York: Semiotext(e).

Index

N

P

Q

R

S